Action and Reaction

Studies on Motion and Contradiction in Physics

Horst-Heino v. Borzeszkowski and Renate Wahsner

Die Deutsche Bibliothek – CIP-Einheitsaufnahme

Borzeszkowski, Horst-Heino v.:
Action and reaction : studies on motion and contradiction in physics / Horst-Heino v. Borzeszkowski and Renate Wahsner. - Berlin : Logos-Verl., 2001

ISBN 3-89722-740-1

ISBN 3-89722-740-1

Logos Verlag Berlin
Comeniushof, Gubener Str. 47,
10243 Berlin
Tel.: +49 030 42 85 10 90
Fax: +49 030 42 85 10 92
INTERNET: http://www.logos-verlag.de

Contents

Introduction

Recognizing that philosophy and the various specialized sciences must work together is often confused with failing to think the difference between philosophy and science.[1] There is a widespread misunderstanding, for example, that output from the sciences in itself constitutes philosophy. The epistemological source of this error is an ontologization of concepts created within these sciences: means generated in order to perceive the nature of actuality are mistaken for the nature of actuality.[2] By blurring the said difference – between physics and philosophy, for example, or between natural science and the dialectics of nature – one resolves neither the problems confronted by specialized science, nor the problems confronted by philosophy. Instead, one produces metaphysics or metaphysicizing empiricism in the sense of Hegel's definitions.[3] Neither of these philosophical standpoints is able to perceive the epistemological status of empirical natural science. As a result, they attribute a role to physics which cannot be played by a measuring, calculating natural science, expecting it to cast philosophical illumination upon the world and, by the same token, not noticing what physics actually accomplishes in observing the self-motion of natural bodies.

As was shown earlier, the epistemological status of physics as a measuring and calculating natural science is characterized by the duality principles on which it rests.[4] In

---

1 What is really meant is the distinction, not the separation of physics and philosophy, not the separating, the *singularization* of differences. The term "distinct" is used in the present paper in the essential differentiation from the term "separate".

2 Cf., E. Husserl, *Die Krisis der europäischen Wissenschaften und die transzendentale Phänomenologie*, in *Husserliana. Gesammelte Werke*, ed. by E. Bienel, Den Haag 1962, p. 52.

3 According to Hegel, metaphysics is the first position of thought to objectivity. It is that innocent procedure which, still unaware of the opposites of thinking in itself and against itself, entails the belief that truth is perceived by mere thinking, that what objects truly are is presented to consciousness by mere thinking. Considering a thing metaphysically means thinking dialectical antitheses externalized against each other. (Cf., G.W.F. Hegel, *Enzyklopädie der philosophische Wissenschaften im Grundrisse. (1830). Erster Teil. Die Wissenschaft der Logik, mit den mündlichen Zusätzen*, in *Werke in 20 Bdn., auf der Grundlage der Werke von 1832-1845 neu edierte Ausgabe*, Frankfurt a.M. 1986, vol. 8, pp. 93-106 (§§ 26-36).) – Empiricism is the second position of thought to objectivity. While it boasts the advantage of rendering concrete contents in the abstract theories of reason, it is marked above all by its unconscious use of the categories. (Ibid., pp. 106-112 (§§ 37-39).) Hegel describes the approach adopted by the sciences as metaphysicizing empiricism. (Cf. G.W.F. Hegel, *Vorlesungen über die Geschichte der Philosophie III*, in *Werke*, op. cit., vol. 20, pp. 209, 223.)

4 Cf., e.g., H.-H. v. Borzeszkowski and R. Wahsner, *Newton und Voltaire. Zur Begründung und Interpretation der klassischen Mechanik*, Berlin 1980; R. Wahsner, *Das Aktive und das Passive. Zur erkenntnistheoretischen Begründung der Physik durch den Atomismus – dargestellt an Newton und Kant*, Berlin 1981; H.-H. v. Borzeszkowski and R. Wahsner, *Physikalischer Dualismus und dialektischer Widerspruch. Studien zum physikalischen Bewegungsbegriff*, Darmstadt 1989; "Quantum Mechanics and the Physical Reality Concept," *Found. Phys.*, vol. 18 (1988), 669-681; "Heisenberg's Cut and the Limitations of its Movability," *Ann. Phys. (Leipzig)*, vol. 45 (1988), 522-528; *Die Wirklichkeit der Physik. Studien zu Idealität und Realität in einer messenden Wissenschaft*, Frankfurt a.M.–Berlin–Berne–NewYork–Paris–Vienna 1992; *Messung als Begründung oder Vermittlung? Ein Briefwechsel mit Paul Lorenzen über Protophysik und ein paar andere Dinge*. Herausgegeben und geführt von Horst-Heino v. Borzeszkowski und Renate Wahsner, Sankt Augustin 1995; *Voltaire, Elemente der Philosophie Newtons/ Verteidigung des Newtonianismus/ Die Metaphysik des Neuton,*

other words, one pillar incorporates epistemological and measurement-theoretical principles, and essentially it constitutes a precondition for whichever theory of physics we care to address, placing it in an *a priori* relationship. The other pillar incorporates those laws of dynamic interaction which explain the physical object concerned. A metaphysical approach to physics ultimately always refuses to recognize that physics needs *both*. The empiricist approach primarily overlooks the significance of the former principles, while the rationalist primarily overlooks the latter. Asserting that physics contains a more or less *a priori* element is not the same as advocating philosophical apriorism. Rather, it is implicitly stating that "physics is not philosophy" and acknowledging thereby that philosophy has a task all of its own. (At this point we will borrow some terminology devised by Newton and draw a distinction by calling the former "passive principles" and the latter "active principles".[5]) Whichever part of physics – "aprioric" or "dynamic" – one chooses to abstract from, one does not ask in either case *how* scientific knowledge is acquired. Essentially this leads one to equate physical theory with actuality and so, inevitably, to negate the *epistemological* issue underlying physics. By adopting this approach, one denies philosophy its right to exist, because if one equates theory with actuality, or thinking and perceiving with natural states, rather than solving the *problem* posed by the relationship betwPlatozed right back to antiquity, one is declaring that the question is pointless.

The issues raised by Kant – "How is science possible?", "How are necessary lawful statements possible?"[6] – articulated the perception of a problem, and this was the foundation for classical German philosophy. The problem cannot be vanished out of existence: it needs resolving, not in the primary interests of a meaningful philosophy of science, but in the interests of philosophy in general, in the interests of contemporary systematic philosophy. But at least two things are essential if one wants to reach beyond that supreme response from classical philosophy: (i) a reasonable reconstruction of the conceptions of natural science that one finds in Kant, Schelling, Fichte and Hegel, and (ii) a reading or processing of modern natural science based on a knowledge and critical appropriation of classical German philosophy. It cannot be left to a few specialists to provide a philosophical reading of the various sciences (even

---

ed. and inroduced by Renate Wahsner and Horst-Heino v. Borzeszkowski, Berlin 1997; "Not Even Classical Mechanics is Mechanistic," in *Worldviews and the Problems of Synthesis,* vol. 4: The Yellow Book of "Einstein Meets Magritte", ed. by D. Aerts, H. van Belle, and J. van der Veken, Dordrecht–Boston–London 1999, pp. 251-262; "Dualism of Physical Space-Time and Dynamics," in *Proceedings of the Conference "Physical Interpretation of Relativity Theory", 11-14th September 1998, London*, ed. by M.C. Duffy, London 1998, pp. 56-62; "Non-locality versus Locality: The Epistemological Background of the Einstein-Bohr Debate," in *Classical and Quantum Nonlocality,* ed. by P. G. Bergmann, V. de Sabbata, J. N. Goldberg. The Science and Culture Series – Physics, Series Editor: A. Zichichi, Singapore–NewJersey–London–HonKong 2000, pp. 21-50.

5 Cf., I. Newton, 31th Question to his *Opticks*. In more detail on the characterization of the active and passive principles, see the literature cited in note 4.

6 See, I. Kant, *Kritik der reinen Vernunft,* in *Immanuel Kant, Werke in 12 Bdn.*, ed. by W. Weischedel, Frankfurt a.M. 1968, vols. III and IV; *Metaphysische Anfangsgründe der Naturwissenschaft,* in *Werke,* op. cit., vol. IX.

though knowledge of the scientific specialization is crucial). This is something which concerns philosophy as a whole.

At the same time, as the above-mentioned research demonstrated, the core questions of basic physics are related to the philosophical problem of how humans acquire scientific insight. The debate about problems which have emerged in the context of modern physics has shown with particular clarity that to acquire knowledge one requires not only a *subject of cognition* and an *object of cognition*, but also a *means of cognition*. The need for a physical means of cognition (a measuring device and the measurement-theoretical presumptions on which it is constructed) is expressed in the very fact that any physical theory requires *a priori* measurement-theoretical presumptions which – as all developments to date indicate – can never be abolished without trace, but only subjected to variations. Epistemological explorations of how physics functions have ultimately also helped to acquire deeper insight into the relationship between actuality and consciousness. Physics is like a model which enables one to study what *scientific* experience is.[7] The quality of the answer influences how deeply one grasps the basic philosophical problem. The great minds of classical German philosophy were presumably aware of this. They integrated a discussion of the experiential sciences into their system, not in order to demonstrate their universal knowledge, but in order to answer the question how and why science is possible, or how one can move from a mere thought to the real world. The answer determined their philosophical system.

Today, regrettably, the metatheory which seeks to grasp science usually excludes philosophy, while philosophy is conducted without a knowledge or understanding of natural science. This could trigger a relapse which throws behind classical German philosophy. Instead, one should *constructive-critically* ensure its acquisition. This should provide a basis for comprehending the status of scientific experience, for grasping the concept of *scientific experience*, without seeking to link the experiential sciences to philosophy simply by means of a methodological settlement, as Kant did, or to dissolve the natural sciences in philosophy, like Hegel. To implement the cognitive options or philosophical approaches which have developed in the last 200 years, it is crucial to recognize Kant's apriorism as an alienated acceptance of the need for a means of cognition which provides the bridge between the subject and object of cognition, and to see the purpose of this confrontation with apriorism as being to cast philosophical light on the nature of the means (and not simply to discard this means).

---

7 This proposition does not mean that science should be reduced to physics. It certainly does not mean that biology, physiology, psychology etc. are ultimately only divisions of physics, particularly as the view represented here that the immediate objects of physics are physical quantities and the relations between them rule out suggesting that the object domain of physics must include everything studied by the natural sciences. However, universal statements about scientific experience must take into account what is explored in physics using the principle of scientific experience. In addition, of course, one should not lose sight of the fact that physics is the most sophisticated natural science to the extent that only physics has managed so far to permit the measurement and calculation of motion.

The present work is an attempt to seek this illumination by investigating the way physics as an experiential science grasps motion conceptually. Of course, *motion* is a theme with an appeal all of its own, in that the birth of natural science in the modern-age *(neuzeitlich)* sense was associated with a radical shift in categories in the wake of which motion was promoted from a predicate to a subject in itself.[8]
Therefore, since the foundation of theoretical natural science, it has been an uncontested fact that the essential subject of physics is the motion.[9] Of course, motion is considered here not only as spatial motion, i.e., as change of place in the time, but as the change (or change of change) of various quantities characterizing physical systems.

As dialectical philosophy determines motion, however, as the existing contradiction, the question arises: How can the contradiction be the subject of physics being a measuring and calculating science and how can physics conceive this subject? The scope of this question is evident when one thinks of the determination of motion by Hegel and his followers. In *The Science of Logic*, Hegel said: "Some thing moves only not when it is in this Now here and in another Now there, but when it is in this Now here and not here, when it is and is not in this Here at the same time. We have to admit that the old dialecticans were right in indicating contradictions in the motion; however, it does not follow from it that the motion is not but rather that the motion is the existing contradiction itself."[10]

---

8 On the problems of this categorical turn, see, e.g., K. Laßwitz, *Geschichte der Atomistik*, Hamburg–Berlin 1890; E. Cassirer, *Substanzbegriff und Funktionsbegriff. Untersuchungen über die Grundfragen der Erkenntniskritik*, Berlin 1910; *Das Erkenntnisproblem in der Philosophie und Wissenschaft der neueren Zeit*, Darmstadt 1994, vols. I-IV, especially vol. I, pp. 18-61; E.J. Dijksterhuis, *Die Mechanisierung des Weltbilds,* Berlin–Göttingen–Heidelberg 1956; M. Wolff, *Geschichte der Impetustheorie. Untersuchungen zum Ursprung der klassischen Mechanik*, Frankfurt a.M. 1978; B. Heidtmann, "Die sich selbst bewegende Substanz. Zu Voraussetzungen und Konsequenzen des philosophischen Grundsatzprogramms Hegels," in *Arbeit und Reflexion. Zur materialistischen Theorie der Dialektik – Perspektiven der Hegelschen Logik*, ed. by P. Furth, Cologne 1980; R. Wahsner, *Mensch und Kosmos. Die copernicanische Wende*, Berlin 1978, especially pp. 6-70, 95-359; *Zur Kritik der Hegelschen Naturphilosophie. Über ihren Sinn im Lichte der heutigen Naturerkenntnis*, Hegeliana. Studien und Quellen zu Hegel und zum Hegelianismus, Bd. 7, Frankfurt a.M.–Berlin–Berne–NewYork–Paris–Vienna 1996, pp. 11-19, 54-60, 116-121, 217-221; H.-H. v. Borzeszkowski and R. Wahsner, "Die Natur technisch denken? Zur Synthese von τέχνη und φύσις in der Newtonschen Mechanik oder das Verhältnis von praktischer und theoretischer Mechanik in Newtons Physik," in *On the Culture of Morality. Practical Philosophy in the Posttraditional Society*, ed. by M. Weingarten, Berlin (in print) [also: Preprint 87 of the Max Planck Institute for the History of Science, Berlin 1998, especially pp. 9-12, 19-26].

9 To our mind, motion is *the* subject of physics. This thesis must be examined, of course. For the purpose of the present paper, however, this is not necessary; it suffices to proceed from the fact that physics deals with the motion. The thesis saying that motion is the subject of physics is not to pretend that physics is the only science to deal with motion.

10 "Es bewegt sich etwas nur, nicht indem es in diesem Jetzt hier ist und in einem anderen Jetzt dort, sondern in dem es in einem und demselben Jetzt hier und nicht hier, indem es in diesem Hier zugleich ist und nicht ist. Man muß den alten Dialektikern die Widersprüche zugeben, die sie in der Bewegung aufzeigen; aber daraus folgt nicht, daß darum die Bewegung nicht ist, sondern vielmehr, daß die Bewegung der daseiende Widerspruch selbst ist." (G.W.F. Hegel, *Wissenschaft der Logik. Zweiter Teil*, in *Werke,* op. cit., vol. 6, p. 76.)

This statement made by Hegel is undoubtedly of philosophical nature and thus it presupposes the *philosophical* concept of motion and body. As it is an uncontested fact that physics succeeds, otherwise, in conceiving the motion scientifically, the statement made by Hegel – provided that it is true – must also be a meaningful statement when it is based on the *physical* concepts of motion and body.

This assumption is contested frequently. Those who makes seriously a difference between physics and philosophy[11] contest the assumption mentioned above because they suppose that physics does not conceive motion and body as a unity but that this unity would be separated into body without motion (motionless body) and motion without body (unbodily motion, i.e., *ungegenständliche Bewegung*).[12] This problematic opinion results from the correct consideration that it is only objects separated from each other that can be determined by comparison, that their momenta must be separated when one aims to measure the motion as it is done in physics. This brings about the conclusion in some cases that modern-age (*neuzeitlich*) science is indifferent against the philosophical problem of motion.[13] Another opinion says that Hegel is completely disproved by set theory and that this theory established proof of the fact that the philosophical paradoxes – the Zeno's paradoxes of motion, above all – were paralogism. Arguing in this sense, Bertrand Russell said: "When a body moves, one can only make the statement that it is in this place at this time and in that place at that time."[14] This anti-dialectical polemic was, however, refuted already at the beginning of our century by the argument that in such a way one does not grasp the motion itself but one describes its result only, that one takes the motion as a sum of states of rest so that the contradiction being the motion is not abolished but only hidden.[15]

In the following, we want to show that physics really conceives the motion as a contradiction, as a dialectical contradiction although in a specific manner owing to its specific character as a measuring and calculating science. We have called this particu-

---

[11] See note 1.

[12] Cf., e.g., B. Hartmann. "Bewegung und Widerspruch. Zu den weltanschaulichen Voraussetzungen des Engelsschen Urteils über den Widerspruch in der einfachen Ortsbewegung," *Dt. Zs. f. Philosophie*, vol. 28 (1980), 848-858.

[13] See, for example, A. Grünbaum, *Philosophical Problems of Space and Time*, Dordrecht–Boston 1973, pp. 158-176.

[14] In detail he writes: "People used to think that when a thing changes, it must be in a state of change, and that when a thing moves, it is in a state of motion. This is now known to be a mistake. When a body moves, all that can be said is that it is in one place at one time and in another at another. We must not say that it will be in a neighboring place at the next instant, since there is no next instant. Philosophers often tell us that when a body is in motion, it changes its position within the instant. To this view Zeno long ago made the fatal retort that every body always is where it is; but a retort so simple and brief was not the kind to which philosophers are accustomed to give weight, and they have continued down to our day to repeat the same phrases which roused the Eleatic's destructive ardor ... But it was obvious that there were infinities ... The contradictions of infinity seemed unavoidable, and philosophy seemed to have wandered into a 'cul-de-sac'. This difficulty led to Kant's antinomies, and hence, more or less indirectly, to much of Hegel's dialectic method." (B. Russell, *Mysticism and Logic and other Essays*, London 1921, pp. 84 f.)

[15] A survey on the discussion about the quoted determination of motion by Hegel is given by M. Wolff in his book *Der Begriff des Widerspruchs. Eine Studie zur Dialektik Kants und Hegels*, Meisenheim 1981, especially pp. 25-36.

larity *physical dualism.*[16] It is reflected in the passive and active principles which are the parts of physics, as described above.

One focus of this work is to explain the "trick" which physics devised in order to grasp motion, described philosophically by Hegel as existing contradiction, in such a way that it could be measured and calculated. With this aim in mind, the dualism of space-time and matter within classical mechanics, general theory of relativity, quantum mechanics and thermodynamics determining its specific measurement-theoretical significance is analyzed. This is critically compared with the manner in which the same dualism finds reflection in the protophysical context of a constructive philosophy of science.

[16] "Dualism" is a term we use to designate the specific grasping or the specific appearance of the (dialectical) contradiction in physics, that is, in an empirical mathematized science. The *dualism* dealt with here has nothing to do with dualism in logic or mathematics or with the understanding-metaphysics that is called "dualism" by Hegel. The existence of dualism in physics is expressed by the fact, for example, that cannonically conjugate quantities occur in classical mechanics and in the classical field theories, and complementary quantities occur in quantum mechanics.

1. The Differential Calculus and the Logically Contradiction-free Formulation of Contradiction

Objections against the ability to conceive physics as the existing contradiction result essentially from the opinion that the statement "... is at a certain time in this place and in another place" brought about, in terms of physical and mathematical considerations, a logical contradiction because it meant that the place of the body at the time t is x and not x, that is, the same quantity had two different values. This misunderstanding results from the belief that physics would determine the state of motion of a body at the time t only by its spatial coordinate x. This is not correct, however, because physics describes the state of motion of a body at the time t by two quantities, that is, the spatial coordinate x and the velocity v. The formulation "... is and is not ..." refers to two different quantities, where the velocity (at the time t at the place x) actually implies a reference to places not equal to x. This does not result in a logical contradiction because there are two different quantities that are algebraically independent from each other and that can be assigned to the same body at the same time point t.

As to rectilinear uniform motion, it is well-known that the quantity *velocity* is defined by establishing a relationship between the difference in time $\left(\frac{x'-x}{t'-t}\right)$; that is, one refers to two different places. Historically seen, it was difficult to find a method with which a certain velocity can be assigned to a certain place in case of a non-uniform motion. The infinitesimal calculus provides this method by forming the limit $\lim_{t'\to t}\frac{x'-x}{t'-t}$ from the difference quotient $\frac{x'-x}{t'-t}$, where the time t' approaches the time t and the corresponding place x' the place x. The result of this mathematical operation is a new quantity, the velocity, that can be and must be assigned to a certain time t and thus to the corresponding place x. Thus, this description of motion is made by using only the quantities *space* and *time*, and this is possible because a dualization in space-time and velocity can be established by means of the differential calculus. At the same time, the kind of establishing the quantity *velocity* clearly shows that this quantity always contains a reference to (infinitesimally) neighbored places.

Even when taking only the purely kinematic aspect into account, the motion is, therefore, not described as the mere sequence of punctual events or positions of rest but, by means of the infinitesimal calculus, it is seen as a contradiction in the sense of the determination by Hegel, of course, in a form that is typical for the mathematical and physical approach. Thus, without creating logical contradictions one can demonstrate the thesis, that a body in motion is and is not at the same time in one place, in such a manner that makes calculation and measurement possible. However, to achieve this it had been necessary to assign the two momenta "is in one place" and "is in another place" to different quantities that belong together but are independent of each other mathematically. The principle developed here for the case of spatial motion (e.g., the change of place with time) has a general meaning for physics: it is the method to describe also all the other physical types of motion, for instance, changes of thermodynamic functions of state with the time or changes of physical fields with place and time.

Such a possibility to grasp motion and thus the contradiction physically, that is, to shape the principle developed by classical atomism (saying how the motion can be conceived without logical contradiction) in such a manner that motion is not only thinkable but also measurable and calculable, was only given with the foundation of the infinitesimal calculus. When mathematics had developed this calculus, that is, when it incorporated the infinite that is nothing quantitative and that was not part of it before, its hitherto existing border was passed and it became relevant for physics.

Hegel reflected the great importance of the infinitesimal calculus philosophically.[17] In this calculus, he writes, mathematics operates with qualitative forms of quantities and thus goes beyond the usual comprehension "by requiring to think the quantum by being infinite as a superseded as such that is not a quantum and whose quantitative determination, however, remains".[18] Hegel states that this mathematical infinite so arises from the concept of the true infinite. He stresses that the infinite quantum is not a determination of quantity that would have an existence as a quantum but that has only importance with regard to something that stands in relationship to it and that is zero outside this relationship. In other words: Only the *relationship* of the infinite, indefinite quanta dx and dy, that is the differential quotient dx/dy is something determined. Both the limit of spatial differences and the limit of time differences alone is zero. The infinite quantum is thus determined only in the essential unity with its counterpart; thus it is the determination of this quantity in a qualitative form. The following objection was made against the differential calculus: An intermediate position between being and nothing is not possible; but the calculus mentioned above is based on this assumption if the calculus proceeds from the fact that those determinations of quantum were vanishing quantities, that is, those that are not any quantum but neither nothing, but a *determination against others, against each other*. This objection was rejected by Hegel who said that the unity of being and nothing is not a state but the disappearing as well as the becoming, that is, the middle or the unity itself, are their truth alone. "Thus, those infinite quantities are not only comparable but they are only as momenta of comparison, of the relationship."[19]

Hegel highly appreciated the method of the infinitesimal calculus as a procedure that corresponds to the true infinite, however, he considered it a shortage that mathe-

---

17 See G.W.F. Hegel, *Wissenschaft der Logik. First Part*, in *Werke*, op. cit., vol. 5, pp. 279-357. In more detail on this question, see: R. Wahsner, "'Der Gedanke kann nicht richtiger bestimmt werden, als Newton ihn gegeben hat.' Das mathematisch Unendliche und der Newtonsche Bewegungsbegriff im Lichte des begriffslogischen Zusammenhangs von Quantität und Qualität," in *Hegels Seinslogik – Interpretation und Perspektiven,* ed. by A. Arndt and Ch. Iber, Berlin 2000, pp. 271-300, and the literature cited therein; H.-H. v. Borzeszkowski and R. Wahsner, "Infinitesimalkalkül und neuzeitlicher Bewegungsbegriff oder Prozeß als Größe," Preprint 165 of the Max Planck Institute for the History of Science, Berlin 2001.

18 "... indem das Quantum, indem es unendlich ist, als ein Aufgehobenes zu denken gefordert wird, das nicht ein Quantum ist und dessen *quantitative Bestimmung doch bleibt.*" (G.W.F. Hegel, *Wissenschaft der Logik. First Part*, op. cit., p. 283.)

19 "Jene unendlichen Größen sind daher nicht nur vergleichbar, sondern sind nur als Momente der Vergleichung, des Verhältnisses." (Ibid., p. 297.)

matics did not deduce this method from the Notion "but that it is visible that the method of the differential calculus was not invented and established for itself"[20] but had been deduced from empirical sentences and justified its value only by the success achieved by means of this calculus. He saw the reason for this shortcoming in the fact that mathematics contradicts with the infinitesimal calculus its own method on which it is based as science. Basically, it is the science of operations with finite quantities; but calculation with the infinite requires procedures that are rejected in the first case; at the same time, the same procedure used in connection with finite quantities are to be used in connection with infinite quantities. This notionless form of the differential calculus in mathematics would imply the incapability of mathematics to prove the qualitative momenta and thus it would be impossible to make this calculus relevant for physics.

Hegel is fully right when he writes: "Mathematics cannot, however, prove determinations of physical quantities as these are laws that are based on the *qualitative nature* of momenta", and he is fully right in saying that the qualitative is outside the mathematical sphere unless taken from experience lemmatically. But he is wrong in saying that this is so "because this science is not philosophy, does not proceed from Notion".[21] The vision of Hegel was a *philosophical* science of mathematics proceeding from Notion that can prove the qualitative momenta and thus overcomes the incapacity of ordinary mathematics. As the teaching of quantities it would be the science of measures[22] and thus provide a "mathematics of nature"[23] that would be – one can suspect – the ideal physics according to the ideas of Hegel. This view prevents to recognize the actual importance of the infinitesimal calculus for the physical grasping of the motion.

Hegel called it preparatory what had be done at this time for a "science of measures". It would be of great benefit to know the "empirical numbers of nature" and, as it was done by Kepler and Galilei, to give them a general form of determination of quantum. But then it would be necessary to recognize these laws from interrelated qualities (that is in case of Kepler's and Galilei's laws from space and time). However, this work had still to be done.

Thus, Hegel recognized that the infinitesimal calculus goes beyond the borders of mathematics counting and considering relations of numbers, but he made the critical remark that there is still something missing that would make it a science. One recognizes that mathematics cannot supply this missing thing. This would be the moment to conceive the independence (*Eigenständigkeit*) of physics and, at the same time, its inseparable connection to mathematics (and philosophy).[24] However, Hegel sees this

---

20 "... daß man es der Methode des Differentialkalküls wohl sogleich ansieht, daß sie nicht für sich selbst erfunden und aufgestellt worden ist; ... ." (Ibid., p. 323.)

21 "... weil diese Wissenschaft nicht Philosophie ist, nicht vom Begriffe ausgeht ... ." (Ibid., p. 321.)

22 Cf., G.W.F. Hegel, *Enzyklopädie der philosophische Wissenschaften im Grundrisse. (1830). Zweiter Teil. Die Naturphilosophie, mit den mündlichen Zusätzen*, in *Werke*, op. cit., vol. 9, p. 54 (§ 259 R).

23 Cf., G.W.F. Hegel, *Wissenschaft der Logik. First Part*, op. cit., p. 406.

24 That what is Hegel fully right in recognizing as a shortage of mathematics, that is, mathematics is not able to deduce the qualitative momenta, is introduced by physics in the form of empirical con-

cognitive finding only in a negative sense by pointing out that the differential calculus provides possibilities reaching beyond mathematics but that is not yet philosophy. This view is fully consistent in his system where there is a too small room for an independent physics working as a measuring natural science with correct independent methods – that is methods that are not contestable philosophically.[25] The consequence of this view, however is that only a directly philosophical grasping of the existing contradiction, of the motion, can be thought. In such a manner, the infinitesimal calculus as a mathematical means to grasp the motion physically is not taken into consideration.

Doing so, Hegel takes the differential quotient dx/dy always for itself, and this only with regard to the infinite; he ignores the fact that a new quantity is created by it and thus the possibility is given – by the dualization in space-time and velocity – to represent physically that "to move means ...: to be in this place and not to be there at the same time".[26]

Of course, Hegel knew that in mechanics the infinitesimal calculus is applied to the representation of the motion. But he considered this fact as a mere application that contaminates the idea of that calculus. Therefore, he appreciated Newton as a mathematician highly but did not understand his physical work and said that Newton did mathematically what Kepler had found by experience.[27] In the above-mentioned remark about the infinity of the quantum, Hegel said: "The thought cannot be determined more correctly than it was given by *Newton.* I separate here the determination belonging to the ideas of the motion and velocity ... because the thought in them does not appear in the appropriate abstraction but concrete, mixed with non-essential forms."[28] After comprehensive studies, he came to the conclusion that the applications of the differential calculus to the "elementary equations of motion" do not offer

---

stants, axioms arising from experience and rules of measurement, in addition to the mathematical calculus. Of course, Hegel would not be satisfied with this procedure. But for a measuring science that needs fixed determinations, another procedure is not possibly; it is not able to develop notions in the sense of Hegel, it can only stipulate developments of notions of philosophy on a certain level of development.

[25] See R. Wahsner, "Naturwissenschaft zwischen Verstand und Vernunft," in *Vom Mute des Erkennens. Beiträge zur Philosophie G.W.F. Hegels,* ed. by M. Buhr and T.I. Oiserman, Berlin 1981; "Die Auswirkung des Hegelschen Physikkonzepts auf die Naturphilosophie der Folgezeit," in *Hegel-Jahrbuch 1991*, ed. by H. Kimmerle, W. Lefèvre and R. W. Meyer, Fernwald (Annerod) 1991; "The Philosophical Background to Hegel's Criticism of Newton," in *Hegel and Newtonianism*, International Archives of the History of Ideas, vol. 136, ed. by M.J. Petry, Dordrecht 1993 [also: Appendix to the present volume, pp. 143-152.; *Zur Kritik der Hegelschen Naturphilosophie*, op. cit; H.-H. v. Borzeszkowski, "Hegel's Interpretation of Classical Mechanics," in *Hegel and Newtonianism*, op. cit. [also: Appendix to the present volume, pp. 135-141]

[26] "Bewegen heißt aber: an diesem Ort sein und zugleich nicht; ... ." (G.W.F. Hegel, *Vorlesungen über die Geschichte der Philosophie I* in *Werke*, op. cit., vol. 18, p. 314.)

[27] See G.W.F. Hegel, *Wissenschaft der Logik. First Part*, op. cit., pp. 320 f.

[28] "Der Gedanke kann nicht richtiger bestimmt werden, als *Newton* ihn gegeben hat. Ich trenne dabei die Bestimmungen ab, die der Vorstellung der Bewegung und der Geschwindigkeit angehören ..., weil der Gedanke hierin nicht in der gehörigen Abstraktion, sondern konkret, vermischt mit außerwesentlichen Formen erscheint." (Ibid., p. 298.)

any "*real*" interest in itself.[29] Thus, it was not by accident that in *The Science of Logic*, he devoted many pages to the book *Theory of Analytic Functions* by Lagrange but he did never mention the main work of Lagrange *Analytic Mechanics.* But the latter was of decisive importance for the elaboration of classical mechanics and the further development of physics because it gave, on the basis of the differential calculus, of the variation calculus and of the d'Alembert principle, a form to Newtonian mechanics that made it possible to apply this theory to solid bodies and liquids and to use it later also for non-mechanical physical systems like, for example, electrodynamic systems. For Hegel, the equations of mechanical motion in the form of Lagrange were, of course, of no interest like these equations in the Newtonian version.

The opinion of Hegel saying that application to the motion would soil the pure thought of the differential calculus was right only to such an extend that at this time this calculus could be justified only with reference to the intuition (*Anschauung*) but not in a purely mathematical sense. One can fully agree with him when he rejected to found the calculus with the "increase in gravity"[30] or the argument of the "unimportance of the difference"[31]. In the time after Hegel, however, the formalism of the infinitesimal calculus was given its justification in the mathematical concept as it was founded arithmetically by Cauchy, Weierstraß, Cantor, Dedekind and others.[32] The foundation consists in the fact that the continuity of real numbers is proved either by establishing proof of the fact that every Dedekind section in the field of real numbers has a number of intersection (Dedekind) or that every fundamental sequence has a limit (Cantor). By reducing the differential calculus to purely arithmetic concepts it was shown that operations that are made by mathematics in the differential calculus do not contradict the nature of finite determinations and their relations, thus mathematics,[33] provided that, in the sense of Hegel, finite determinations are taken as thought determinations that remain in opposition to each other as well as against the objective and the absolute.[34]

But such a foundation by the Notion was not in the sense of Hegel. This is evident because it does not give the "mathematics of nature" as he had desired. Hegel also gave the reason for it: Mathematics can take over the qualitative only from experience but cannot develop it from it self. However, he was of the opinion that the philosophical Notion is able to do so. Hegel ignored that the philosophical Notion is able to do so only on condition of a principle of scientific experience.[35] Thus, he believed that

---

29 "Für sich bietet so die Anwendung des Differentialkalküls auf die elementarischen Gleichungen der Bewegung kein *reelles* Interesse dar." (Ibid., p. 348.)

30 Cf. ibid., p. 319.

31 Cf. ibid., pp. 301 f., 307 f.

32 See, for example, B.R. Dedekind, *Stetigkeit und irrationale Zahlen*, Berlin 1965.

33 As to this criterion given by Hegel, see G.W.F. Hegel, *Wissenschaft der Logik. First Part*, op. cit., pp. 296 f.

34 Cf., G.W.F. Hegel, *Enzyklopädie der philosophische Wissenschaften im Grundrisse (1830). Erster Teil. Die Wissenschaft der Logik*, op. cit., p. 91 (§ 25).

35 As to the distinction between scientific experience and experience gained in the everyday life, see K. Bayertz, "Zum Verhältnis von erkenntnistheoretischen und ontologischen Aspekten der materialistischen Naturdialektik," in *Theorie und Labor. Dialektik als Programm der Naturwissenschaften,*

an unmediated synthesis of mathematics and philosophy, a "philosophical mathematics" was possible and brought about a "science of measures".

Hegel's critique of the calculus is rooted partly in the inadequate mathematical foundations from which it suffered in his day, and partly in a conceptual misunderstanding which reveals Hegel's limited view of *function.*[36] Due to a peculiar differential procedure, Hegel interprets the transition from an original function $f(x)$ to the derived function $f'(x)$ or to the differential quotient dx/dy as the transition from a determined quantitative relationship, an *equation*, to a *ratio* in the form of a fraction interpreted on the basis of conceptual logic (and no longer an equation).[37]

This limited concept of function is inherent to Hegel's view of the relationship between the differential calculus and physics in the sense of formalism and content. Although Hegel insists that the infinitesimal calculus is founded on mechanics,[38] on the other hand he regards the "derivation" or proof of the gravitation law simply as a *formal generalization.*[39] He feels that differential calculus as it has been developed is being exploited to pretend that physical connections of a problematic nature have been proven, and even that purely formal connections which are only admissible in mathematics are being presented as physical evidence. "It is presented as a triumph of science that by means of mere calculus and *beyond the scope of experience* laws, i.e., sentences of existence, are discovered which do not exist."[40] Kepler, he argues, has already established everything there is to establish about the physical nature of the gravitation law.[41] This is an old view of Hegel's, and one to which he was evidently deeply attached, for it tinged his concept of heaviness and gravitation and his understanding of the relationship between mathematics and natural science as long as he lived. The literature has expounded frequently on why it was a false view.[42] Above all, Hegel overlooks the character of physical dynamics, as opposed to the pure kinematic

---

ed. by P. Plath and H.J. Sandkühler, Cologne 1978, pp. 37-50; H.-H. v. Borzeszkowski and R. Wahsner, "Zur Beziehung von experimenteller Methode und Raumbegriff," *Dt. Zs. f. Philosophie*, vol. 28 (1980), 685-696; "Physikalische Erfahrung und physikalische Wirklichkeit," in *Die Wirklichkeit der Physik*, op. cit., pp. 239-285.

36 Cf., R. Wahsner, "'Der Gedanke kann nicht richtiger bestimmt werden, als Newton ihn gegeben hat'," op. cit.; H.-H. v. Borzeszkowski and R. Wahsner, "Infinitesimalkalkül und neuzeitlicher Bewegungsbegriff oder Prozeß als Größe," op- cit.

37 Cf., G.W.F. Hegel, *Wissenschaft der Logik. First Part*, op. cit., pp. 335-345, 350.

38 Cf. ibid., e.g., pp. 322-358, especially 346-348.

39 Cf. ibid., pp. 320-322.

40 "Es wird für einen Triumph der Wissenschaft ausgegeben, durch den bloßen Kalkül *über die Erfahrung hinaus* Gesetze, d.i. Sätze der Existenz, die keine Existenz haben, zu finden." (Cf. ibid., p. 320.)

41 Cf., G.W.F. Hegel, *Enzyklopädie der philosophischen Wissenschaften im Grundrisse (1830). Zweiter Teil. Die Naturphilosophie*, op. cit, vol. 9, p. 87 (§ 270 R).

42 Cf., e.g., H.-H. v. Borzeszkowski, "Hegel's Interpretation of Classical Mechanics," op. cit.; R. Wahsner, "The Philosophical Background to Hegel's Criticism of Newton," op. cit.

theory of Kepler's astronomy, and the radical change in thinking which was needed in order to come up with a theory of motion such as Newton's.[43]

As a result of the conceptual-logical development in the *qualitative quantity ratio, quantity* reverts to *quality*. Accordingly, Hegel regards the physical significance of the separate terms in a series formed, according to Lagrange, by differentiation, or in the physical substantiation for interrupting the series, as a procedure which might go some way towards meeting his purposes. He defines the so-called applications as the thing itself[44] or else comments "that surely one can see at once from the method of the differential calculus that it was not invented and established for its own sake".[45] Hegel did not intend to rest content with merely superficial "application" or assigning the mathematical calculus to "physical being". He wanted a solid theory to substantiate the connection. As Hegel realized, mathematics was unable to provide one. It could only deliver a formal-universal. His hopes for a "mathematics of nature" were informed, therefore, by a recognition that there were limits to naked mathematics and that a description of nature needed to build on a cohesive theory.

Hegel states on a number of occasions that mathematics was overstepping its limits in developing this calculus and including the infinite, which is not quantitative and therefore not within its former domain. He does not recognize, however, that this is what made it relevant to physics as a *theory of motion*. It was basically his concept of differentiation which prevented him from noticing this potential. Physical motion is described by whole equations, if not systems of equations, and not by individual terms, even if these are "power relationships" (*Potenzverhältnisse*). Hegel's method does not permit the idea of a change in the change (e.g., in velocity with respect to time, where velocity is itself a rate of change in distance with respect to time) – and via the concept of acceleration this is after all quite essential to the physical view of motion – because the only way to obtain such a concept is to differentiate the same function twice in its entirety. But that means, the first derivative does not change an equation to a mere ratio, it provides again an equation. The derivative from this derivative must be $f''(x)$ the second derivative from the original function $f(x)$.

To grasp motion as a contradiction, we need more than two elements in the relationship. (In the simplest (kinematic) case one needs place, time and velocity, the third of these itself reflecting a relationship between the first two. A "point" relationship – such as the fraction interpreted in conceptual logic as the relationship between *number* and *unity* and between an *extensive* and *intensive* quantity – cannot do that. In fact, the *relationship* is the whole "equation" – in its entirety, including the original function and its derivatives. the infinitesimal calculus cannot be dealt with in terms of the quantum category. Hegel was obliged to re-interpret Newton's idea of fluxion in

---

43 Some aspects of this turn are considered in: E. Cassirer, *Substanzbegriff und Funktionsbegriff*, op. cit., e.g., pp. 88-99, 173-184; see also: H.-H. v. Borzeszkowski and R. Wahsner, "Die Natur technisch denken?, "op. cit., especially pp. 9-12, 19-26.

44 Cf., G.W.F. Hegel, *Wissenschaft der Logik. Erster Teil*, op. cit., pp. 325, 335, see also 309 f.

45 "... daß man es der Methode des Differentialkalküls wohl sogleich ansieht, daß sie nicht für sich selbst erfunden und aufgestellt worden ist". (Ibid., p. 323.)

order to express this as an infinite quantum.[46] Hegel had his reasons for including the *contradiction* which establishes its existence as motion, not within his teachings on being, but within his teachings on essence. The modern-age *(neuzeitlich)* concept of function can only be grasped under the category *relation*, but in the Kantian tradition (not to mention older ones) categories of quantity are different from those of relation. the infinitesimal calculus required a new categorical context to reveal that this calculus gave the mathematical precondition for making measurable the unity of discontinuity and continuity of space with the continuity and discontinuity of time, a unity as which he determined the motion in his philosophical analysis of eleatic paradoxes.

In discussing those paradoxes, Hegel recognizes that the thinking is forced to hold the momenta of an object (*Gegenstand*), that are connected in actuality, in their difference. From this necessity results the difficulty of thinking.[47] And he believes that these momenta can be brought together again only by means of philosophy. No doubt about, this is true, but precisely taken only because valid forms of this interconnecting already exist on the level of fully developed special sciences and because philosophy depends on these sciences as far as it cannot mediate thinking and being without these sciences.

Of course, the infinitesimal calculus alone cannot link thinking and being, it does not suffice to measure the motion, it *per se* cannot describe motion physically. For studying the specifically physical grasping of contradiction, it is necessary to see physics as a whole, to consider physics as dynamics in its epistemological foundation. Before doing so it is necessary to make a further remark about the infinitesimal calculus.

---

[46] See also W. Bonsiepen, "Hegels Theorie des qualitativen Quantitätsverhältnisses," in *Konzepte des mathematisch Unendlichen im 19. Jahrhundert*, ed. by G. König, Göttingen 1990, p. 125.

[47] Cf., G.W.F. Hegel, *Vorlesungen über die Geschichte der Philosophie I*, in *Werke*, op. cit., vol. 18, p. 314.

2. Analytical Geometry as the Principle of Linking Mathematics and Measurement

The contradiction being the motion can be represented by means of the infinitesimal calculus; this does not result, however, from this mathematical calculus for itself but from the fact that physical determinations were assumed right from the beginning, that is, the entities in the calculus were identified with physical quantities of measurement, that is, with differences in space and time and with velocities. This identification is only possible by linking geometry and algebra, that is, on the basis of analytical geometry.[48, 49]

To explain it, we shall first confine our discussion the three-dimensional geometry of space. On the one hand, analytical geometry permits to correlate different dimensions (lengths, widths and heights, say Δx, Δy and Δz) and to use them for calculation according to the rules of the algebra of finite and infinitesimal quantities. Assuming that Δx, Δy and Δz can be measured, the combinations established by these quantities algebraically $\Delta x \cdot \Delta y$, $\Delta x/\Delta y$, dx/dy etc. are quantities of measurement. Thus, it is possible to form equations by using these quantities, equations that can be of physical meaning. On the other hand, analytical geometry makes it possible to represent algebraic equations or their solutions in a purely geometric form, namely, as curves on which lie the points determined by the equations. This equivalence between algebra and geometry does not only reveal – as pointed out frequently – the existence of general (abstract) mathematical structures but it renders it possible to establish a *mathematicized measuring* science. This is true because they are quantities of measurement only in case of the mapping of numbers Δx, Δy and Δz to spatial differences of geometry, of the Euclidean geometry to use a more precise term. This is based on the fact that the Euclidean geometry is the theory of rigid bodies and their possibilities of displacement and that only rigid bodies are used as measuring bodies. Via the analytical Euclidean geometry, that is, via the equivalence of algebra or arithmetics and Euclidean geometry, other analytical geometries may be important physically as far as the respective geometries are provided with certain characteristics of the Euclidean geometry (one example is the Riemann geometry in the theory of general relativity) or as they can be related to it (as, for example, in the quantum mechanics) in such a manner that measurement is possible.[50] Thus, analytical geometry gives a principle for relating mathematical relations to reality.

---

[48] Sometimes, one finds in the literature the standpoint that the field of modern mathematical physics entered into by Varignon and Lagrange, and that Euclidean geometry, no longer being essential, was abandoned. (Cf., e.g., M. Blay, *Reasoning with the Infinite. From the Closed World to the Mathematical Universe*, Chicago–London 1998.) This opinion oversees completely that analytical geometry is algebrized ("analytized") *Euclidean* geometry and that only this geometry enabled one to found the differential calculus and, as described in this chapter, to relate mathematical relations to reality.

[49] For the history of analytical geometry, see, e.g., C. B. Boyer, *History of Analytic Geometry*, NewYork 1956.

[50] The foundation of the Gauß-Riemann geometry showed that only specific properties of the Euclidean geometry are required for making the measurement possible. Helmholtz was the first to

If one wants to calculate and measure the *motion*, time to has be included, of course. The algebraic constructions mentioned above can be extended respectively when one takes a general approach to what was said about analytical geometry and when one takes into account not only lengths, widths and heights (let us say Δx, Δy and Δz again) but also time in the sense of a dimension –[51] and this means, according to the founder of analytical geometry, a determination and relation in which a object can be measured and considered similar to Δx, Δy and Δz.[52] This results in the new terms $\Delta x/\Delta t$, dx/dt etc. that can be measured in principle. However, the Cartesian scheme does not determine *how* the time is to be measured and *how* it is to be linked with spatial differences. The answer to this question is identical with the development of physics as the science of motion and thus with the determination of a standard motion (see Sections 3 to 5).[53]

The analytical geometry had a physical aim right from the beginning owing to the fact that it was developed by Descartes to create the possibility to measure natural motions. His aim was "to direct the human cognitive power to the creation of unshakable and true judgments on every thing what exists".[54] He held the view that this can be attained by elaborating his "Mathesis universalis". This he considered as the general

---

make this finding. According to his view, it is necessary to use a geometry in which the bodies can be moved without changing their forms and their volums. Thus, he also admits non-Euclidean spaces if a rigid body can be moved in them as freely (without deformation) as in the Euclidean space, that is, all spaces with a constant curvature, not only the special case of a space with the curvature zero. (See H. v. Helmholtz, "Ueber den Ursprung und die Bedeutung der geometrischen Axiome," in *Vorträge und Reden von Hermann von Helmholtz*, vol. II, Braunschweig 1903, pp. 10 f.) The foundation of the theory of general relativity that is, the identification of gravitational field and curvature (metrics) showed that spaces of arbitrary curvature changing from one world point to the other must be admitted. Nevertheless, the demand for the possibility to measure is satisfied by the fact that Riemannian geometry is quasi-Euclidean (Minkowskian) in the infinitely small and makes a procedure possible that can be used for a comparison of distances on various space-time points.

51 This is proved convincingly by d'Alembert in the introduction to his *Traité de Dynamique*. (See the quotation in note 156.)

52 Descartes says: "As to dimension, we understand nothing else but the determination and relation in which a thing is considered to be measurable so that not only length, width and depth are dimensions of a body but also the weight a dimension according to which things are weighted, velocity a dimension of motion and an infinitely great number of similar things. ["Per dimensionem nihil aliud intelligimus, quam modum et rationem, secundum quam aliquod subiectum consideratur esse mensurabile, adeo ut non solum longitudo, latitudo, et profundidas sint dimensiones corporis, sed etiam gravitas sit dimensio, secundum quam subiecta ponderentur, celeritas sit dimensio, motus , et alia eius modi infinita," (R. Descartes, *Regulae ad directionem ingenii*, Leipzig 1907, p. 53.)]

53 See, for example, J.B. d'Alembert, *Traité de dynamique dans lequel les lois de l'équilibre et mouvement des Corps sont reduites au plus petit nombre possible, et demontrés dune maniere nouvelle, et ou l'on donne un Principe général pour trouver le Mouvement de Corps qui agissent les uns sur les autres d'une maniere quelconque*, Paris 1976, p. 11.; C. Neumann, *Über die Principien der Galilei-Newtonschen Theorie*, Leipzig 1870. – With reference to it, Cassirer finds out: " The law of inertia is incorporated into the explanation of the time unit as an integral conceptional part." ["Das Trägheitsgesetz geht als begrifflicher Bestand in die Erklärung der Zeiteinheit ein." (E. Cassirer, *Substanzbegriff und Funktionsbegriff*, op. cit., p. 192.)]

54 "Studiorum finis esse debet ingenii directio ad soloda et vera, de iris omnibus quae occurunt, proferenda iudicia." (R. Descartes, *Regulae ad directionem ingenii*, op. cit., p. 3.)

science of order and measure where it is not important "whether such a measure is to be found in numbers, figures, stars, sounds or any other objects".[55] He believed to develop a comprehensive theory of nature by means of this mathematics of abstract structures hidden behind various fields of arithmetics, geometry, astronomy, theory of music, optics and mechanics.[56]

The elaboration of this "Mathesis universalis" was identical with the foundation of analytical geometry. For Descartes, this was more than the application of algebraic methods for solving geometric problems or the geometric illustration of given equations and their solutions (as it was seen by Fermat). For him, this was the mathematical-constructive part of his cognitive method. Therefore, he did not found it within a special mathematical work. It is rather – rule 18 is one example – the result of his epistemological considerations in the *Rules Relating to the Adjustment of the Cognitive Power* (*Regulae ad directionem ingenii*).

According to these considerations, Descartes related geometry not only to the spatial dimension, but for him geometry was a model that made it possible to deduce mathematically by using all auxiliary sources of the intellect, the power of imagination, the senses and the memory.[57] The geometry as taken by Descartes in his *Rules* as an analytic one right from the beginning could provide this model only when he left geometry dimensionally undetermined. First he did not determine either the number of dimensions (the number of axes of Cartesian system of coordinates) or the determined nature of these dimensions. And thus the uniform distances on the axes).[58] When founding his method, his aim was only to achieve that various determinations of nature can be seen from the point of view of analytical geometry so that they can be used

---

55 "Quod attentius consideranti tandem innotuit, illa omnia tantum, in quibus ordo vel mensura examinandur, ad Mathesim referri, nec interesse utrum in numeris, vel figuris, vel astris, vel sonis, aliove quovis obiecto talis mensura querenda sit; ac proinde generalem quamdam esse debere scientiam, quae id omne, explicet, quod circa ordinem et mensuram nulli speciali materiae addicta quaeri potest, eamdemque, non ascitiv vocabulo, sed iam inveterato atque usu recepto, Mathesim universalem nominari, quoniam in hac continetur illud omne, propter quod aliae scientiae et Mathematical partes appelantur." (Ibid., p. 13.)

56 Descartes thought that they had already known this true mathematics in ancient times. He wrote: "I think the traces of this mathematics being true up to the present time were already visible in the works of Pappus and Diophant. ... However, this teaching was suppressed later by the writers with a fatal slyness." ["Et quidem huius verae Mathesa vestigia quedem adhuc apparere mihi videntur in Pappo et Diophanto, qui, licet non prima aetate, multis tamen saeculis ante haec tempora viscerunt. Hanc vero postea ab ipsis Scriptoribus perniciosa quadam astutia suppressam fuisse crediderim, nam sicut multos artifices de suis inventis fecisse compertum est, timnerunt forte, quia facillima erat et simplex, ne vulgata vilesceret, malueruntque nobis in eius locum steriles quasdam veritates ex consequentibus actulus demonstratas, tanquam artis suae effectus, ut illos miraremur, exhibare, quam artem ipsam docere, quae plane admirationem sustulisset." (Ibid., p. 12.)]

57 "Denique omnobus utendum est intellectus, imiaginationes, sensus et memoriae auxilis, tum ad propositiones simplices distincte intuendas, tum ad quaesita cum cognitis rite componenda ut agnoscantur, tum ad illa invenienda, quae ita inter se debeant conferri, ut nulla pars industriae humanae omittatur." (Ibid., p. 32, see also pp. 32-44.)

58 Of course, dimension does not mean, as applied by Descartes, the same as, say, in the theory of special relativity of today. According to Descartes, this concept refers to the fact that x and t can be measured and combined algebraically and not to the fact that they must be strongly correlated – like in the theory of special relativity – with regard to the invariance group of the theory of motion.

for calculation despite different dimensions. Which dimensions are "real" is more a matter for the teachers in natural science, Cartesius said.[59]

The great importance of the Cartesian analytical geometry and thus of all geometries that are sufficiently similar to the geometry of Euclid results from the fact that it makes it possible to link mathematics and measurement. Descartes' identification of calculating quantities and Euclidean geometric quantities became decisive for physics when Euclidean geometry became part of the measurement-theoretical basis of physics. That identification of calculating and geometric quantities established a connection between calculating quantities and quantities of measurement. (As indicated above, this does not mean, of course, that all quantities in a physical theory must be quantities of measurement.) In this context, it was of specific importance that the Cartesian concept of dimension later has allowed to include the time, as mentioned above. Descartes was not able, however, to determine the actually fundamental dimensions in his "Physics" that is, to say which quantities of measurement are to be attributed to the various axes. Thus, he identified the dimensions with the three spatial ones and tried to deduce all the other dimensions from them.

Owing to the interconnections mentioned above, it did not happen by chance that motion became calculable by means of the infinitesimal calculus. It did not happen by chance because that calculus was developed on the basis of analytical geometry. The mathematicized natural science "physics" did not come into being, since the mathematical calculus was conceived and interpreted physically later but the source of this linkage between mathematics and natural science, a linkage being extremely close and seeming nearly "prestabilized harmonically", is analytical geometry.

Thus, one can fully agree with the view that the scientific discovery of analytical geometry is connected with a genuine philosophical revolution in thinking.[60] This is so, as Cassirer wrote, because the logic of space is transformed into logic of number, because space was brought to an exact conceptualization by impressing the same epistemological character to it like it was given to numbers so far and thus transforming the mathematical functional connection to a "logic of manifolds" reaching beyond the limits of syllogistics.[61] However, the revolution of the form of thinking owing to analytical geometry is fully conceived only when one recognizes that its essence also lies in providing a principle according to which mathematical relations can be related to reality and the epistemological character of space can be impressed the numbers.

---

59 "Ex quibus patet, infinitas esse posse in eodem subiecto dimensiones diversas, illaque nihil prorsus superaddere rebus dimensis, sed eodem modo intelligi, sive habeant fundamentum reale in ipsis subiectis, sive ex arbitrio mentis nostrae fuerint excogitate: est enim aliquid reale gravitas corporis, vel celeritasmotus, vel divisio saeculi in annos et dies; non autem divisio diei in horas et momenta, etc., quae tamen omnia eodem se habent modo, si considerentur tantum sub ratione dimensionis, ut hic et in Mathematicis disciplinis est faciendum: pertinent enim magis ad Physicos examinare, utrum illarum fundamentum sit reale." (Ibid., p. 53.)

60 Cf., E. Cassirer, *Substanzbegriff und Funktionsbegriff*, op. cit., p. 93.

61 Cf. ibid.

Then, mathematics for itself makes always statements on abstract-possible relations alone but not on real-possible ones.[62]

A true relationship between reality and its abstract reflections by means of a mathematized theory can only be mediated by physical basic principles or basic principles of other special sciences. Their importance is not seen frequently due to the fact that it is possible successfully to use specific rules of calculations for the implementation of certain tasks. This success allows the conclusion that reality was reflected correctly but it is, however, only based on the fact that basic physical principles were simplified and changed into what we call rules of calculation. The conditions allowing the respective simplification is taken implicitly. Abstracting from them basically, physical principles are degraded to rules of thumb de facto or rules of thumb are given the values of basic physical principles. If these procedures are used in technics, the appearance of a purely mathematical reflection of reality can go to pieces rather clearly.[63]

62 This fact implies the problems that one encounters in the foundation of geometry. For a representation of different approaches to geometry, see: P. Mittelstaedt, Philosophische Probleme der modernen Physik, Mannheim–Wien–Zürich 1989, pp. 49-73.

63 Cf., K. Zweiling, *Gleichgewicht und Stabilität*, Berlin 1953. On page 13 he writes, for example: "In this field [of the use of physical theories in engineering technology] there had been an theoretical chaos. ... Most superficial empirism developed obtrusively. Empirists known as 'important figures in theory' indulged in trivial falsifications of theory. Second-class and third-class rules of calculation – simplified for the every-day domestic use from actual basic principles, useful only for urgent every-day work – were made equal to theoretically substantiated basic principles arbitrarily. Thus, the basic principles were degraded to rough and ready formulas and reversely, rough and ready formulas were given the status of basic principles." [Auf diesem Gebiet [der ingenieur-technischen Anwendung physikalischer Theorien] herrschte damals ein theoretisches Chaos. ... Der oberflächlichste Empirismus machte sich aufdringlich breit. Als 'theoretische Größen' anerkannt Kritiker gefielen sich in flachen Verfälschungen der Theorie. Zweit- und drittrangige Rechenregeln – aus wirklichen Grundprinzipien für den alltäglichen Hausgebrauch simplifiziert, in sehr beschränkten Bereichen durchaus nützlich für eilige Alltagsarbeit – wurden den theoretisch einwandfrei fundierten Grundprinzipien willkürlich gleichgestellt. Damit wurden auch die Grundprinzipien zu Faustformeln degradiert und umgekehrt die Faustformeln zum Rang von Grundprinzipien erhoben."]

3. The Dualism of Physical Space-time and Dynamics as a Consequence of the Possibility to Measure and Calculate Motion

The principle of physics, to connect the momenta which one had distributed at first, the principle of physics to think the motion, was established by classical atomism and perfected in the following centuries for the purpose of physics.[64] The philosophical system of Greek atomism developed for the first time the principle of a physical representation of nature. This principle bases on the atomistic main idea to reduce the whole world, all existing, to the atom and the void, to the full and the empty, to being and non-being, to body and space. By this reduction it was established the conception of the actuality of entities being opposed to each other: The being was presupposed as real as the non-being, discontinuity as real as continuity, rest as real as motion. Both momenta are to exist really but they must have existences that are differentiated from each other and that exist as two determined beings that are separated from each other and belong together at the same time. This distribution of conceptual momenta existing only in unity, through the distribution of opposite determinations and their reformulation into relatively independent entities made it possible to grasp the motion in such a manner that contradiction, constituting the motion, could be thought free from logical contradiction.

Their philosophical system was the atomists' reaction to the recognition, arrived at by predecessors, the Eleatics and the Heracliteans, that motion cannot be conceived without contradiction and therefore, as they concluded, cannot be conceived at all. This recognition took form in two opposing viewpoints. The Eleatics considered that in truth the world was motionless and comprehensible, conceivable, just because of motionlessness. In the eyes of the Heracliteans, on the other hand, the world is in permanent motion, but precisely for that reason it cannot be known or conceived but can at most be adequately reflected in sensual perception. Motion seemed unthinkable both in rationalist-Eleatic and in empirist-Heraclitean conception, because not-being was seen as not thinkable. For what was – like not-being – not an object, what thus has no specificable features, cannot be conceived in thought. Consequently the idea that something is and is not, that something by being in motion is at one place and simultaneously is not there, is likewise taken to be an unthinkable contradiction. The atomists solved the problem of motion, of conception and of actuality which their predecessors had posed from opposing viewpoints – a problem which sprang from the insight that the thinking (always in some way discontinous) and real existence (in the last analysis always continous) cannot directly correspond. Thus, the atomists solved a problem springing from the fundamental question in philosophy which here was explicitly posed for the first time.

The ancient atomists found a hihgly ingenious escape from this dilemma: by means of their reduction of everything that exists to atoms in a void they exhibited the possibility of conceiving not only what is but also what is not, so that being and not-being can be thought of as coexisting in the world. Atoms represent being, the void repre-

---

[64] See R. Wahsner, *Das Aktive und das Passive*, op. cit., especially pp. 26-32.

sents not-being, and one is as real as the other. The actuality of what is not splits up what is, so that the pieces of what is may now have different and contrary descriptions. What is is continous and discontinous, indivisible and divisible, constant and variable, infinite and finite. (And it is no longer impossible to think that something, by being in motion, both is and not is.) The first term in each pair concerns the individual atom, the atom by itself, the second concerns compositions of atoms, that is to say macroscopic bodies and more in general atoms in relation among each other. That it is the same being to which these different descriptions apply is guaranteed by the fact that all atoms are qualitatively equal: they all have only size and form, the differences among them are merely quantitative.

This distribution of moments for the first in the history of human thought permitted discriminating between matter and space. It also made it possible to grasp motion conceptually, in such a fashion that the contradiction constituting motion can be thought physically, i.e., by making motion measurable and calculable. Without the concept of space the simultaneous actuality of being and not-being could not have been maintained. But now space appeared as the condition for making motion possible and therefore also as the condition for the possibility of comprehending, of thinking about the world – such that logical contradictions do not arise.

To recognize that concept and actuality are not identical and to recognize the contradictoriness of motion thus formed the condition for developing the physical thought-principle The atomists' solution for making motion thinkable and bringing thought and actuality into coincidence was one possible solution of this problem of motion, thought and actuality. It was that particular solution that took nature as its undoubted presupposition. It conceived the entire universe as nature; consciousness, the subject of cognition, functions merely as an external observer.[65] This is just the point of view of science. The other possible (and indeed necessary) solution of the problem is the philosophical one, embodied in the birth of Plato's dialectics.

The atomist conception made motion physically thinkable; but it could not establish a necessary connection between this thought-principle and sensual perception. In other words, while ancient atomism developed the principle of physical thinking, it could not established a principle of physical sensibility. For this purpose atomism had to be modified.

In the original atomist view the atom is in essence postulated as a single, unrelated individual. But making the particle's individuality absolute in this way cannot provide a sufficient gnoseological basis for physics, at least not in a way that would link theory and measurement. The limit of the atomist view appears in its attempt to explain the diversity of the world, the genesis and decline of nature, by means of properties that may be ascribed to a body, to an atom as a single.

---

[65] See Schrödinger's principle of objectivization. (Cf. E. Schrödinger, *Nature and the Greeks*, in *Nature and the Greeks/ Science and Humanism*, with a foreword by R. Penrose, Cambridge 1996; H.-H. v. Borzeszkowski and R. Wahsner, "Non-locality versus Locality: The Epistemological Background of the Einstein-Bohr Debate," op. cit.; R. Wahsner, "Schrödinger's Reception of Greek Atomism," in *The Concept of Probability*, ed by E.I. Bitsakis and C.A. Nicolaides, Dordrecht 1989.)

The isolation of the physical body, conceived as an atom, later was eliminated by the physical (not the mechanistic) concept of force, by the laws of physical interactions. The physical concept of force comprises the idea that natural bodies act on each other, that in referring or better said relating to each other they exert a force, and therefore are not something strictly isolated and essentially singular.

The atomistic approach, of course, brings about metaphysical mechanicism if actuality is assumed to be so *per se*. If it is taken as a method, however, which does not operate with real entities but with thought entities (physical quantities, for example), this approach founds the principle of physical thinking.

The characteristic of this above-mentioned principle implies that dualisms occur in physics necessarily: the dualism of the space-time structure and physical interactions, that is, of geochronometry and dynamics, of particles and field, of passive and active principles, of topology and metrics etc.[66] In these cases, the two sides of a dualism within the framework of a respective physical theory are not on the same footing but the first side bears the character of a precondition related to the basic equations of the respective theory. This different weighting of the two sides was not yet given by classical atomism because classical atomism made the physical motion thinkable but could not establish a necessary connection between the principle of thinking as developed by it and the sensual perception because it could not say how to *produce* thought entities, measuring quantities. The different weighting of the two poles of dualism as appearing in (developed) physics is conditioned by the requirements of measurement and calculation, by the physical method to link sensuality and understanding.

A difference in weighting – although physically not capable of bearing – can already be found in the phenomenological "physics" of Aristotle who made a difference between natural and forced motions. The way towards physics as an empirical mathematized natural science can also be characterized as endeavor to find out what has to be determined as the presupposed side of dualism and what as the side that has to be explained physically. Galilei was the first to go the physically constructive step in this direction when he pre supposed the rectilinear uniform motion as the basic property of bodies. On the basis of making a difference between rectilinear uniform motion and motions deviating from them Newton was able to develop the classical mechanics where, by means of physical dynamics, only motions that deviate from rectilinear uniform motions are explained.

As physics conceives the contradiction constituting the motion by the above-mentioned distribution of momenta, by the distribution of opposite conceptual determinations it must make a certain divide between the two sides of the dualisms. The history of the development of physics has shown that it is of importance how this divide between the existing-in-unity momenta of real motion is made.

---

66 This point of view was founded in: H.-H. v. Borzeszkowski and R. Wahsner, "Über die Notwendigkeit der Philosophie für die Naturwissenschaft," in *Dialektik 1. Beiträge zu Philosophie und Wissenschaften. Orientierungen der Philosophie*, ed. by B. Heidtmann, Cologne 1980; it was elaborated in *Physikalischer Dualismus und dialektischer Widerspruch*, op. cit.

For instance, the duality of natural and forced motions – the conception developed by Aristotle and having its effects until the time of Renaissance – proved to be inadequate for being used as the conceptual basis of physics. As the concept "natural motion" was determined on the basis of a geocentric point of view and was deduced from the material nature of bodies, it was not possible to determine it in a uniform manner for all motions in nature: a difference had been made between supralunear and sublunear (and in the latter between falling and rising) natural motions. Thus, physics as a science of the whole nature could not be established, even not a physical dynamics of the terrestrial nature.[67]

But even a heliocentrically oriented distribution of momenta abstracting from the material nature of bodies and avoiding a breaking of the world into two fields is physically not capable of bearing if, like in the conception of Kepler, for all motions existing in the world the rest is determined as the moment which must be presupposed.[68] Kepler said that if there was only one body in the world, it would be in rest because rest was the basic property of the matter; if a second body was brought into the world, however, these two bodies bring each other into motion. As a conclusion for the system of planets he wrote that as all matter tends to rest in a place where it is (unless attracted by the magnetic power of a neighboring body), it happens that the moving power of the sun gets into conflict with this inertia of the matter, like two weights struggle with each other on the balance, and that from the ratio of the two forces it results a higher or lower speed of the planets.[69] Later one succeeded in describing the motion of celestial bodies as a conflict between inertia and gravitation but one succeeded in doing so only by another determination of inertia.[70] The principle of inertia being a fundamental of classical mechanics postulates the rectilinear uniform motion equivalent to rest as the basic property of physical bodies and thus the gap between rest and motion is excluded right from the beginning.[71] In this sense,

67 As to the problems of the conception of peripatetic motion and its reception, see M. Wolff, *Geschichte der Impetustheorie*, op. cit. R. Wahsner, *Mensch und Kosmos*, op. cit., pp. 21-31.

68 As to the contribution of Kepler to the development of the theory of physical motion, see R. Wahsner, "Weltharmonie und Naturgesetz. Zur wissenschaftstheoretischen und wissenschaftshistorischen Bedeutung der Keplerschen Harmonienlehre," in *Dt. Zs. f. Philosophie*, vol. 29 (1981), 531-545; H.-H. v. Borzeszkowski, "Hegel's Interpretation of Classical Mechanics," in *Hegel and Newtonianism*, op. cit.; H.-H. v. Borzeszkowski und R. Wahsner, "Die Natur technisch denken?," op. cit.

69 "Da jegliche Materie zur Ruhe neigt, an dem Ort, an dem sie sich befindet (wenn nicht ein benachbarter Körper sie durch eine magnetische Kraft an sich zieht), geschieht es, daß die bewegende Kraft der Sonne in Widerstreit gerät mit dieser Trägheit der Materie, wie an der Waage zwei Gewichte miteinander streiten, und daß aus dem Verhältnis der beiden Kräfte schließlich eine größere oder geringere Geschwindigkeit der Planeten herauskommt." (J. Kepler, *Mysterium Cosmographicum – Das Weltgeheimnis*, Augsburg 1923, pp. 105 f. – German translation.)

70 Koyré also underlined this point of view strongly when he said: "We see that Kepler, the great Kepler, the founder of modern astronomy, the same man who proclaimed the unity of matter in the whole universe and stated that *ubi materia, ibi geometria*, failed to established the basis of modern physical science for one and only one reason: he still believed that motion is ontologically on a higher level of being than rest." (A. Koyré, "Galilei and the Scientific Revolution of the Seventeenth Century," in *Metaphysics and Measurement*, London 1968, p. 12.)

71 In his Introduction to the *Mathematical Principles*, Newton writes: "... motion and rest alone as they are seen normally, differ from each other by the kind of relation, and those bodies are not al-

d'Alembert writes: "The motion and its general properties are the first and most essential subject of mechanics; this science presupposes the existence of the motion."[72] Thus, mechanics makes the necessary divide not between rest and motion but between rectilinear uniform motion and accelerated motion, divide in the sense that the first is supposed to be a motion etalon and that only motions deviating from it are explained by physical dynamics. – Of course, this determination of a standard motion by Newton is opposite to the basic principle of the general reciprocity of motion as formulated by Galilei and Huygens, supported by Leibniz and discussed by Kant. But the approach of Newton was the only one which made it possible to measure the motion.[73]

The prehistory of physics, seen from this viewpoint, presents itself as endeavor to find how the opposite conceptual momenta really existing in a unity must be distributed in order to make the motion not only thinkable physically but also measurable and calculable. In the following history of physics, one aim was to find out whether the divide can be made in another way (for example, in such a manner that less *"a priori"* determinations were necessary) [74] or whether it can be avoided generally (what is attended by unified theories).

---

ways in rest, that are considered to be in rest normally". ["... sed motus et quies, uti vulgo concipiuntur, respectu solo distinuuntur ab invicem; neque semper vere quiescunt quae vulgo tanquam quiescentia spectantur". (I. Newton, *Philosophiae Naturalis Principia Mathematica*, Glasgue 1822. Vol I, p. 4.)]

72 "Le Mouvement et ses propiétés générales, sont le premier et le principal objectde la Méchanique; cette Science suppose l'exostence du Mouvement ... ." (J.B. d'Alembert, *Traité de Dynamique*, op. cit., p. V.)

73 This statement does not deny the importance of the contributions of other scientists before and after Newton for the development of what one calls today Newtonian mechanics.

74 See H.-H- v. Borzeszkowski and R. Wahsner, "Erkenntnistheoretischer Apriorismus und Einsteins Theorie. Einstein in seiner Beziehung zu Newton und Kant," in *Dt. Zs. f. Philosophie*, vol. 27 (1979), 213-221; "Non-locality versus Locality: The Epistemological Background of the Einstein-Bohr Debate," op. cit.; H.-H- v. Borzeszkowski, "Kantscher Raumbegriff und Einsteins Theorie," *Dt. Zs. f. Philosophie*, vol. 40 (1992), 36-41; R. Wahsner, "Erkenntnistheoretischer Apriorismus und die neuzeitliche Physik," *Dt. Zs. f. Philosophie*, vol. 40 (1992), 24-35; Apriorische Funktion und aposteriorische Herkunft. Hermann von Helmholtz' Untersuchungen zum Erfahrungsstatus der Geometrie, in *Universalgenie Helmholtz. Rückblick nach 100 Jahren*, ed. by L Krüger, Berlin 1994.

4. The Galilean-Newtonian Determination of the Dualism of Space-time and Dynamics – the Basis of Modern Physics

Classical mechanics as the first natural science in its proper sense defines the presupposed "natural" motion by the principle of inertia. In his *Mathematical Principles of the Natural Philosophy* Newton presented this principle as the first law in the following form: "Every body remains in its state of rest or in rectilinear uniform motion unless forced to change its state by acting forces."[75]

The term "natural motion" used for historical reason is not to provoke the idea that this is a motion that can be seen from every-day experience. It is very unlikely to find a rectilinear and uniform motion in nature. To realize it, one has to create ideal conditions artificially (and it is well-known that this can only be achieved approximately).[76] The equivalence of such a motion with rest is also not sensually perceptible but a theoretical conclusion that was only possible after finding out that the rectilinear uniform motion can be defined uniquely by the quantity *velocity*.[77] Compared with this statement, the Aristotelian "natural motion" was far closer to every-day experience. But this was just the reason why it was only possible to establish a phenomenological natural philosophy on this basis, and why these conceptual determinations could not be used for developing a physical theory.[78]

The formulation of the principle of inertia proves that physics does not think the separation of body and motion even under the aspect of force-free motion and does not presuppose an unmoved body or a unbodily (*ungegenständliche*) motion. That assumed separation presupposes that it is reasonable or even necessary in physics to make a difference between a body in rest and a body in motion in such a manner that the motion, as an additional property, is attached to the body definable *per se*. But the principle of inertia excludes a qualitative difference between rest and motion. This

---

[75] "*Lex I.* Corpus omne perseverare in statu suo, quiescendi vel movendi uniformiter in directum nisi quâtenus a viribus impressis cogitur statum illum mutare." (I. Newton, *Philosophiae Naturalis Principia Mathematica*, op. cit., p. 15.)

[76] In recognizing this necessity and the importance of this necessity for the formulation of physical concepts, the constructive philosophy of science in its protophysical form makes a difference between two processes of forming concepts – the *abstraction* and the *ideation*. According to this conception of philosophy of science, concept to be used in the theory of measurement can only be formulated by the latter: "In contrast to the Aristotelian abstraction ... the ideation also contains standards of action, for example for the production of planes and orthogonalities." (P. Lorenzen, "Theorie des technischen Wissens," in *Protophysik. Für und wider eine konstruktive Wissenschaftstheorie*, Frankfurt a.M. 1976, p. 63.) [Although the German term "Wissenschaftstheorie" instead of "Wissenschaftsphilosophie" is more correct, here the usual terms "philosophy of science" and, respectively, "scientific-philosophical" are applied.]

[77] See also P. Jaeglé, who writes: "The velocity can become an essential feature and one of the dimensions of the motion only when it is presented as constantly invariable in the course of time and as preserved in the course of motion. Place and moment are changing but they remain identical as to their intervals irrespective of the form of the latter and how small they are thought. Outside this identity, there is no exact concept of velocity and there is no mechanics." (P. Jaeglé, *Essai sur l'espace et la temps*, Paris 1971, or, in German translation, P. Jaeglé, *Raum und Zeit. Grundfragen naturwissenschaftlicher und philosophischer Erkenntnis*, Cologne 1980, p. 35.)

[78] See note 35.

does not imply, of course, that no difference could be made between rest and motion in physics. A science that is to measure the motion can and must be able to state whether a velocity is equal to zero or different from it. The fact that rest and motion are equated qualitatively and do not represent two different qualities is shown in that they are given by different values of one and the same quantity, that is, of the velocity.

To avoid misunderstandings let us to say that the body is not liquidated in kinematics which studies the motion (the accelerated, too) of physical bodies and systems without taking the moving forces into consideration. However, the body can be characterized more precisely only by the subsequent dynamics for which kinematics is constructed as a presupposition. Seen from a purely kinematic point of view, it occurs only as a moved "something". If one wants to interpret kinematics as a natural theory – as Galilei had done because of the lack of physical dynamics – this will be possible only to the extent of succeeding in replacing the properties of bodies, following from the dynamics in case of an elaborated physics, by additional assumptions. When Newton elaborated the classical mechanics, he presumed the moving "something" as mass – thus anticipating certain results of the dynamics – that can be measured by weight of the respective body as he had found out by his "very precise pendulum tests". Newton designated the resistance against a change in the state of motion as the basic property of this mass. Definition 3 of his *Principles* reads: "Matter is able to resist; therefore, each body, as far as itself it is concerned, remains in a state of rest or in uniform rectilinear motion."[79] This ability proportional to the body or the mass "is different from inertia only as to the approach", Newton said.[80]

As mentioned above, it is necessary to create ideal conditions for producing a uniform rectilinear motion; this means that the respective body has to be isolated from its physical environment so that it cannot be effected by any forces. Since we can suppose that it is possible to realize a uniform rectilinear motion, one has to assume the possibility of isolating physical bodies or systems. Reversibly, the principle of inertia shows what one considers as an isolated body or system in physics.

However, the principle of inertia does not only contain the thought of the ability to isolate but also the necessary reference to a second object. Without such an object, motion cannot be noticed generally. But talking about physics, the object of reference has to satisfy demands: There must not be any interaction between the body in rectilinear uniform motion and the object of reference (as it has to be a force-free moved body), and the object of reference be of such a nature that a rectilinear uniform motion can be noticed in reference to it, that an Euclidean straight line can be defined by it. Such an object of reference is just the homogeneous, infinitely expanded Euclidean space. This space can be considered as a specifically prepared body which makes it possible to define the rectilinear uniform motion for measurement. This necessity was pointed out by d'Alembert: "In order to get a clear idea about motion one is not able to dispense with two space ideas at least in silence. ... One considers the

---

79 "Materiae vis insita est potentia resistendi, qua corpus unumquodque, quantum in se est, perseverat in statu suo vel quiescendi vel movendi uniformiter in directum." (I. Newton, *Philosophiae Naturalis Principia Mathematica*, op. cit., p. 3.)
80 Ibid.

space as impermeable and as the thing what is called the body; the other considers it simply as a spatial extension without investigating the fact whether it is permeable or not, that is as the measure of the distance between one body and the other whose parts conceived solidly and may serve for evaluating rest and motion of bodies."[81]

The fact that, for reasons mentioned above, the principle of inertia contains the Euclidean space as a necessary presupposition clearly shows that physics can think motion only as a correlation of bodies even in the case where the mutual effects of bodies must be excluded; thus the concept of an unbodily motion does not occur in physics even not in connection with the purely kinematic aspect.[82] The object of reference is *quasi* taken as a body without behavior but it is determined as body in such a way that it contains the reference to other bodies implicitly.

Thus, the "natural" motion presupposed by classical mechanics is determined via the principle of inertia; as etalon of motion, it is used to measure accelerated motions deviating from it and being produced by forces. Basically, the sense of physics consists in the lawful representation of these motions by means of dynamics; on the other hand, only dynamics clearly shows the genuine importance of the principle of inertia as described so far. To consider the problem how the above-cited contradiction is grasped through the physical motion, it is indispensable to show how in dynamics in connection with its kinematic presuppositions the relationship between physical body and physical motion is described.

The core of the Newtonian dynamics is the second law saying that the change of the velocity of a body, that is, the acceleration, is proportional to the acting force.[83] To come to a law describing a specific physical interaction, a force term is to be inserted for the force that is limited by axiomatics but not determined by it. For example, when

---

81 "... nuos contenderous de remarquer, que pour avoir une idee claire du Mouvement, on ne peut se dispenser de distinguer au moins par l'esprit deux sortes d'étendue: l'une, qui soit regardée comme impénétrable, et qui constitue ce qu'on appelle proprement les corps; l'autre, qui étant considérée simplement comme étendue, sans examiner si elle est pénétrable ou non, soit la mesure de la disrance d'un corps à un autre, et dont les parties envisagées comme fixes et immobiles, puissent servir à juger du repos ou du mouvement des corps." (J.B. d'Alembert, *Traité de Dynamique*, op. cit., p. VI.) – Here d'Alembert used the term "two space ideas" instead of terms "moved body" and "reference body" used above.

82 In his paper "Galilei and the Scientific Revolution of the Seventeenth Century," A. Koyré comprehensively examined the physical and epistemological presupposition implied by the principle of inertia. However, he misunderstands its measurement-theoretical function. Moreover, he does not see the bodily (*gegenständlic*h) character of measurement. Thus, he understands the Galilean teaching of motion as a victory of Plato over Aristotle. (See A. Koyré, op. cit., pp. 1-15.)

83 Sometimes one meets with the opinion that Newton's second law "is not to be confused with the one expressed in differential terms that we know today as 'Newton's law'". (M. Blay, *Reasoning with the Infinite. From the Closed World to the Mathematical Universe*, Chicago–London 1998, p. 56.) – Of course, in Newton's *Principia*, one does not find this law in an analytic form but in a corresponding geometric formulation. Now and again this fact leads, for example, to the curious misunderstanding that *time* does not occur in this theory. Such a view ignores that Newton defines a *change of motion* as change per time-interval. Possibly, the misunderstanding arises because Newton does not mentioned this explicitely, and for people who do not follow his deductions this might be difficult to see. [For the relation between the geometric and alalytic formulation in Newton, see, e.g., H.-G. Schöpf, "Newton zwischen Geometrie und Erfahrung," in *Gottfried Wilhelm Leibniz im philosophischen Diskurs über Geometrie und Erfahrung*, ed. by H. Hecht, Berlin 1991.]

the force term of gravitational force is used in the second law of Newton, one gets the law of gravitation. This law grasps the gravitational interaction of n bodies in such a way that the acceleration of each of these bodies can be determined through its gravitational interaction with the other $n-1$ bodies. The second Newtonian law just wins a physical sense from the fact that it is clarified in relation to which reference systems the forces and accelerations contained in this law are defined; these are those determined by motions of n bodies if the forces acting between them were excluded (that is, if the bodies were isolated in the sense of the first axiom), and by all motions being rectilinear and uniform with respect to them. The fact that, in case of a given term of force $\mathfrak{F}$, the equation of motion determined by the second law of Newton are valid with respect to systems of reference (inertial systems) defined in this way is expressed in the Galilean principle of relativity.[84] The class of inertial systems defines the physical space (within the framework of classical mechanics) on which, as background, the (accelerated) motion of the physical matter (example is the above-mentioned gravitational interaction of n bodies) is described.[85] Thus, the absolute space of Newton is not a transcendental phantom but a product of idealization that is to determine the standard motion. This function of the absolute space was not recognized for a long time.[86] Even at the end of the last century, it had been necessary to clarify this point.[87]

---

84 See H.-J. Treder, *Über Prinzipien der Dynamik von Einstein, Hertz, Mach und Poincaré*, Berlin 1974, pp. 15-23, 65.

85 On the relationship between Newton's three laws see the editors' annotations and epilogue to: Ernst Mach, *Die Mechanik in ihrer Entwicklung – historisch-kritisch dargestellt*, hg. und mit einem Anhang versehen von R. Wahsner and H.-H. v. Borzeszkowski, Berlin 1988, pp. 534-536, 600-604; *Physikalischer Dualismus und dialektischer Widerspruch*. op. cit., in particular pp. 31-39.

86 Cf., H.-H. v. Borzeszkowski and R. Wahsner, "Mach's Criticism to Newton and Einstein's Reading of Mach: The Stimulating Role of Two Misunderstandings," in *Mach's Principle: From Newton's Bucket to Quantum Gravity*, ed. by J.B. Barbour and H. Pfister, Einstein Studies, vol. 6, Boston–Basel–Berlin 1995.

87 In this context, we want to mentioned especially E. Mach, C. Neumann and L. Lange. (E. Mach, *Die Mechanik in ihrer Entwickelung – historisch-kritisch dargestellt*, Leipzig 1883; C. Neumann, *Über die Principien der Galilei-Newtonschen Theorie*, op. cit.; L. Lange, *Die geschichtliche Entwickelung des Bewegungsbegriffs und ihr voraussichtliches Endergebnis. Ein Beitrag zur historischen Kritik der mechanischen Prinzipien*, Leipzig 1886.)

Lange summarized his thoughts saying that the dynamic (true) motions are determined in relation to the standard motion given by the law of inertia, by stating that this law is adequate to the following definitions and theorems: "*Definition I.* An 'inertial system' is such a system of coordinates that with reference to it the paths (meeting in one point and described continuously) of *three* points simultaneously projected from the same point of space and moving free unmediately (and not lying on the same straight line) are all *rectilinear*. *Theorem I.* In relation to an inertial system, the path of *every fourth* point moving freely is also *uniform*. *Definition II.* An 'inertial time scale' is such a time scale in relation to which any point moving freely is moved uniformly along its inertial path. *Theorem II.* As to an inertial time scale, *each other* point moving freely is moved uniformly along its inertial path." ["*Definition I.* 'Inertialsystem' heißt ein Coordinatensystem der Beschaffenheit, daß mit bezug darauf die in einem Punkte zusammenlaufenden stetig beschriebenen Bahnen *dreier* gleichzeitig von demselben Raumpunkte projizierter und so fort sich selber überlassener Punkte (die aber nicht auf einer Geraden liegen sollen) sämmtlich *geradlinig* sind. *Theorem I.* In Bezug auf ein Inertialsystem ist auch die Bahn eines *jeden vierten* sich selbst über las se nen Punktes *gleich. Defini tion II.* 'Inertialzeitscala' heißt eine solche Zeitscala, in Bezug auf welche irgend ein sich selbst über-

The distribution of momenta of the contradiction that appeared in kinematics as the dualism of space-time and velocity is changed into the dualism of space-time and acceleration by passing over to dynamics so that the kinematic space-time is different from the dynamic space-time due to the different distribution. The new, dynamic dualism makes it possible and requires the introduction of new quantities (mass, charge) that cannot be created by the combination of space and time quantities alone. As to the conceptual basis of physics, the dualism of space-time and matter proved to be decisive, and all the other dualisms or conceptual distribution had to be subordinated to it.

When the interconnection between the contradiction inherent in motion as such and physical motion is considered in philosophical literature the physical reflection of this contradiction is often not seen in the above-described dualism but in the physical interaction (that is, in the allegedly active principles alone), where this interaction is not even taken as the expression of the second Newtonian law but as the equivalence of action and reaction.[88] It is true, this equivalence is a part of classical mechanics but it is not postulated in the second Newtonian axiom but in the third one. It limits (in accordance with the Galilean principle of relativity) the possible force terms which must be inserted in the second law so that the motion can be calculated and physical natural laws can be formulated. The effect of this limitations is that the motions described by the second law actually refer to inertial systems determined in the first law, that is, to the motion-etalon as described here. Thus, the action-reaction law is really important for the Newtonian physics; however, it is not the physical incarnation of the contradiction constituting the motion. This is shown, for example, in the existence of physical theories where forces do not satisfy the action-reaction law. This happens when, like in electrodynamics, for example, the force term contains not only distances, masses and charges but also absolute velocities. Therefore, this law is broken in all special-relativistic field theories. The same objection is valid in principle against all attempts to interpret different physical principles (e.g., the principle

---

lassener Punkt in seiner Inertialbahn gleichförmig bewegt ist. *Theorem II.* Rücksichtlich einer Inertialzeitscala ist auch *jeder andere* sich selbst überlassene Punkt in seiner Inertialbahngleichförmig bewegt." (L. Lange, op. cit., pp. 140-141, see also 54-55.)]

Assuming now that one has a straight line and a clock (geochronometry) and additionally a dynamics, so that one knows what a rectilinear uniform motion or a force-free motion is, the determination made by Lange explain the physical content of the Newtonian absolute space. But it is just the problematic point that cannot be solved from a purely conventionalistic standpoint.

The hope to realize inertial systems by using bodies with a large inertial mass m_I and a small gravitational charge is illusionary, of course, because due to the equivalence of inertial and gravitational masses, very inertial bodies would also have a very masses m_P. Thus, they would be influenced by other bodies gravitationally in an especially strong manner so that this is not force-free motion any more. A quite similar problem exists in the theory of general relativity as to the possibility to measure the gravitation field; there it leads to the Planck's elementary length as a basic measuring border. (See Section 4.2.)

88 The reason for it is to be found in the mechanistic interpretation of the second Newtonian law. If one reads it as a cause-effect relationship in such a manner that force would be the cause of motion or acceleration, one comes only via action-reaction law to an interaction. In reality, however, the second law says that the force $\mathfrak{F}$ is equal to the product $m\mathfrak{b}$, where $\mathfrak{F}$ is only the symbol for the term of force which has still to be introduced. This term substituted for $\mathfrak{F}$, the thus resulting equivalence expresses the respective physical interaction.

of complementary in quantum mechanics) independently from the measurement-theoretical basis of the respective physical theory as interpretational principles at all.

The dualism of space-time and matter, being a basic feature of physics as described above, must be given a new definition by respective physical theory. Classical mechanics determined this dualism for whole physics as far as it occupies a specific position in it, a position that makes it necessary to refer to the basic concepts developed by it again and again.[89] But this does not mean that the border between the two sides of dualism could not be shifted by other physical theories.

It is well-known, for example, that the general relativity theory took the gravitational interaction from the side of dynamics to the side of space-time by the geometrization of gravitation. Thus, the dualism was defined basically newly: the gravitational ether – a term introduced by Einstein and Weyl – replaced the class of inertial systems forming the absolute space; the gravitational ether unifies gravitation and inertia to form a new background on which all non-gravitational interactions are de scribed and measured. The former rigid Newtonian space-time background is changed in a what Einstein called reference mollusc (*Bezugsmolluske*). Otherwise, because rigid measurement rods are necessary for measurement, the theory of general relativity is also forced to refer to a *quasi*-Newtonian situation if it is to establish a connection to the experiment, if it is to establish the possibility of physical interpretation.[90] As long as relatively weak gravitational fields are concerned, this is possible rather easily. However, in case of strong gravitational fields, the modeling of the reference mollusc by means of *quasi*-rigid Newtonian space-time is not possible, even not approximately. The resulting difficulties linked with the revealing of the physical content of general relativity theory are one of the most important topics in discussions dealing with modern gravitational research today. What matters is to establish a connection with classical mechanics for measurement-theoretical reasons and, at the same time, to elaborate a theory with which it is possible to grasp physical situations that differ widely from situations still described by classical mechanics.

Since the geometrization of gravitation also means a physicalization of geometry, as far as space-time structures were identified with the physical field of gravitation, not all space-time determinations are presupposed now but are dynamically determined partly. Thus, the *a priori* measurement-theoretical part of general relativity theory was reduced as compared to that in classical mechanics.[91]

---

89 Cf. e.g., M. Planck, *Das Weltbild der neuen Physik*, Leipzig 1985, p. 20; A. Einstein, "The Mechanics of Newton and Their Influence on the Development of Theoretical Physics," in *A. Einstein, The World as I See It*, translat. by A. Harris, NewYork 1934; A. Einstein, "Isaac Newton," in *A. Einstein, Out of My Later Years*, NewYork 1950; W. Heisenberg, *Wandlungen in den Grundlagen der Naturwissenschaft*, Leipzig 1942; H.-J. Treder, "Das Korrespondenzprinzip und die Bedeutung von Quanten- und Relativitätstheorie," in *Spektrum*, vol. 7 (1976), 37; H.-J. Treder and R. Rompe, *Über Physik. Studien zu ihrer Stellung in Wissenschaft und Gesellschaft*, Berlin 1979, pp. 21, 50, 55, 71; H.-H. v. Borzeszkowski and R. Wahsner, *Newton und Voltaire*, op. cit.; *Messung als Begründung oder Vermittlung?* , op. cit.; "Die Natur technisch denken? ," op. cit.

90 *Quasi*-Newtonian means that all correlations of the Newtonian theory are valid in the first approximation, and the effect of the Einstein theory can be presented as perturbations of this basic situation.

91 For more details, see note 74.

The necessity to determine the dualism of space-time and dynamics in a new manner is especially evident in quantum mechanics.[92] The basic dynamic law of that theory, the Schrödinger equation, can only be formulated in a many-dimensional configuration space which is basically different from the space determined measurement-theoretically. Heisenberg designated the dualism of space-time and dynamics even as a cut. He said: "The situation of any experiment in nuclear physics is as follows: We use more or less complicated apparata for putting questions to nature, questions that are always directed to the definition of any objective process in space and time … . Thus, it is an automatic result that in dealing with the process mathematically, we make a cut between the apparata we use as auxiliary means for putting our questions, on the one hand, … and the physical systems about which we want to learn something, on the other. The latter, we represent mathematically by a wave function and for this wave function a differential equation exists according to the quantum theory which determines the future state of the wave function on the basis of the present one. As to the apparata, however, we content ourselves with laws formulated in classical concepts which provided us with the right to use them as apparata of measurement. The cut between the system to be observed and the measurement apparata is given by our questions naturally but it does not show a discontinuity of the physical process evidently."[93]

To make at all a linkage to experience, that is, to establish a connection between dynamics and measurement determinations, axioms relating to the measurement process are added to the Schrödinger equation. They indicate how to come to the three- or four-dimensional space-time of measurement via the projection of the multi-dimensional configuration space. By this, the gap is closed between the dynamic and the measurement spaces in such a manner that space-time of measurement and the Schrödinger equation do not exist out of context with each other but both of them become part of one and the same physical theory.

Thus, we are dealt here with the case that space in which basic physical equations are formulated does not coincide with the three- or four-dimensional space of experience where measurement apparata are acting. In classical mechanics, the specific case

---

92 Cf., H.-H. v. Borzeszkowski and R. Wahsner, "Quantum Mechanics and the Physical Reality Concept," op. cit.; "Heisenberg's Cut and the Limitations of its Movability," op. cit.; "Non-locality versus Locality: The Epistemological Background of the Einstein-Bohr Debate," op. cit.

93 "Die Situation irgendeines Experiments in der Atomphysik ist die folgende: Wir stellen mit Hilfe von mehr oder weniger komplizierten Apparaten Fragen an die Natur, die stets auf die Feststellung irgendeines objektiven Ablaufs in Raum und Zeit gerichtet sind … . Aus dieser Sachlage ergibt sich automatisch, daß wir bei der mathematischen Behand lung des Vorganges einen Schnitt ziehen einerseits zwischen den Apparaten, die wir als Hilfsmittel zum Stellen der Frage betrachten, … und an dererseits den physikalischen Systemen, über die wir etwas wissen wollen. Die letzteren stellen wir mathematisch durch eine Wellenfunktion dar, wobei für diese Wellenfunktion nach der Quantentheorie eine Differentialgeichung besteht, die den zukünftigen Zustand der Wellenfunktion aus dem gegenwärtigen bestimmt. Für die Apparate dagegen begnügen wir uns mit den in den klassischen Begriffen formulierbaren Gesetzen, die uns das Recht geben, sie eben als Meßapparate zu verwenden. Der Schnitt zwischen dem zu beobachtenden System und den Meßapparaten ist durch unsere Fragestellung naturgemäß gegeben, bezeichnet aber offenbar keine Diskontinuität des physikalischen Geschehens." (W. Heisenberg, "Prinzipielle Fragen der modernen Physik," op. cit., pp. 46-47.)

occurs that dynamics is related to this measurement space (the absolute Newtonian space). As long, as one only knew classical physical theories, the opinion was prevailing that both kinds of space were identical. One knew, of course, that mathematically abstract spaces can be introduced that allow a more elegant formulation of dynamics under specific conditions. However, it was always possible to formulate dynamics relating to the measurement space, too. This is different as to quantum mechanics.

The dualism of space-time and dynamics leads "from the determination of Schrödinger's wave equation in the configuration space to probabilistics in the physical space".[94] The probabilistic aspects of quantum mechanics result from a linkage of the Schrödinger equation (or its relativistic generalization like Klein-Gordon equation, Dirac equation or Fock equation) to measurement. In a lecture given in 1932, Max Planck expressed this situation. He said: "Generally, the laws of material waves are completely different from those inherent in the classical mechanics of material points. What matters, however, is the fact that the function characteristic for material waves, the wave function or probability function (it does not matter which name is used), is completely determined by the initial and boundary conditions for all places and times, according to certain rules of calculation, irrespective of the fact whether the *Schrödinger* operators are used or the *Heisenberg* matrixes or the *Dirac* q-numbers. – We se: determinism governs in the world picture of classical physics, it is only the symbols and stipulations of calculation that are different. According to this, in quantum physics, like in classical physics before, the uncertainty of predictions relating to events in the sensual world is reduced to the uncertainty of the connection between world picture and sensual world, that is, to the uncertainty of translation of world-picture symbols to sensual world, and vice versa."[95]

Thus, it is clearly shown that it is a misinterpretation saying that the so-called statistical causality brings a subjective element into the description of phenomena. This epistemological error – the founders of the quantum mechanics did not make it – re-

---

94 "Die Reduzierung der Freiheitsgrade – vom Konfigurationsraum wird in den dreidimensionalen Raum projiziert – führt von der Determiniertheit der Schrödingerschen Wellengleichung im Konfigurationsraum auf die Probabilistik im physikalischen Raum." (R. Rompe and H.-J. Treder, *Über Physik*, op. cit., p. 69.)

95 "Im allgemeinen sind die Gesetze der Materiewellen grundverschieden von denen der klassischen Mechanik materieller Punkte. Wesentlich ist aber der Umstand, daß die für die Materiewellen charakteristische Funktion: die Wellenfunktion oder Wahrscheinlichkeitsfunktion – der Name tut hier nichts zur Sache – durch die Anfangsbedingungen und die Randbedingungen für alle Orte und Zeiten vollständig determiniert ist, nach ganz bestimmten Rechenregeln, sei es, daß man sich dabei der *Schrödinger*schen Operatoren oder der *Heisenberg*schen Matrizen oder der *Dirac*schen q-Zahlen bedient. – Wir sehen: In dem Weltbild der Quantenphysik herrscht der Determinismus ebenso wie in der klassischen Physik, nur sind die benutzten Symbole andere, und es wird mit anderen Rechnungsvorschriften operiert. Dementsprechend wird in der Quantenphysik, ebenso wie früher in der klassischen Physik, die Unsicherheit in der Voraussage von Ereignissen der Sinnenwelt reduziert auf die Unsicherheit des Zusammenhangs zwischen Weltbild und Sinnenwelt, d.h. auf die Unsicherheit der Übertragung der Symbole des Weltbildes auf die Sinnenwelt, und umgekehrt." (M. Planck, *Der Kausalbegriff in der Physik*, Leipzig 1958, pp. 13 f.) [The fact has to be taken into account that Planck uses the term "sensous world" instead of the terms "space of physical experience" or "space-time of measurement".]

sults from the insufficient analysis of the conditions under which phenomena in physics are studied.[96]

These considerations invalidate arguments saying that quantum mechanics disproves the concept of reality. According to these arguments, the reality concept was said to have been falsified because quantum mechanics had proved the impossibility to describe nature in a causal manner. The scheme of such an argumentation starts from the difference between quantum mechanics and classical mechanics and concludes that, as quantum mechanics would be modern physics at all, the concept of a causality finds no place in modern description of nature and thus the strict lawfulness would be disproved. This approach, however, is an absolutization of classical mechanics by designating its laws alone as adequate to a complete, lawful description of nature. However, this is not the standpoint of physics but of philosophical mechanicism. On the other hand, one underestimates the importance of classical mechanics by ignoring the necessary linkage of quantum mechanics, expressed through the Bohr principle of complementary and of correspondence (Ehrenfest theorems), to classical mechanics; but only this linkage makes quantum mechanics physics. So to say – as Rosenfeld say – an epistemological error!

Thus, Einstein's discontentedness with quantum mechanics did not result from the fact that he considered this physical theory as being indeterministic (what he did not do). He rejected the attempts of Bohm and de Broglie as too simple attempts to describe the quantum mechanics determinsistically[97] and he denied having used the question "Is the theory strictly deterministic?" as a criterion for an admissible theory.[98] As he said, his concern was the so-called real description. Einstein regarded the quantum mechanics as an incomplete and indirect description of reality, which would be replaced by a complete and direct one later.[99] He hoped that the development of physics would result in a theory in which measurement space and representation space would coincide. Einstein's effort aiming at achieving a real description did not arise from a naive philosophical conception, a mechanistic one, for example, but from a profound physical program. His aim was to link modern physics to classical physics without any sharp cuts by formulating the dynamics again with an immediately reference to the experience space.[100]

All discussions about quantum mechanics are concentrated on the topic to solve the problem of the cut between systems to be studied and the measurement apparatus,

---

[96] Cf., C. George, I. Prigogine, and L. Rosenfeld, "The Macroscopic Level of Quantum Mechanics," in *Selected Papers of Léon Rosenfeld,* ed. by R.S. Cohen and J. Stachel, Dordrecht–Boston–London 1979, p. 593.

[97] Cf. A. Einstein to M. Born, letter of 12 May, 1952, in *Albert Einstein – Hedwig und Max Born. Briefwechsel 1916-1955,* commented by M. Born, München 1969, p. 258.

[98] Cf., W. Pauli to M. Born, letter of 31 March, 1954, in *Albert Einstein – Hedwig und Max Born. Briefwechsel 1916-1955,* op. cit., p. 293. – In this letter Pauli brings the dispute between Born and Einstein to a solution.

[99] Cf. A. Einstein to M. Born, letter of 5 April, 1948, in *Albert Einstein – Hedwig und Max Born. Briefwechsel 1916-1955,* op. cit., p. 234.

[100] Compared with the discussion between Einstein and Bohr, Boltzmann had used quite analogous arguments against the mechanics elaborated by Heinrich Hertz. (See R. Rompe and H.-J. Treder, *Über Physik,* op. cit., pp. 25-29.)

a problem occurred in quantum mechanics *explicitly* for the first time. Answers have to found to such questions like, for example, why does the cut exist, where has it to be made and must it occur in the preciseness given by quantum mechanics. As this cut is due to different spaces (dynamic space and experience space), the attempt to unsharpen it results in a revision of the respective space concepts. (Discontinuities occurring in the projection of representation space to the measurement space, for example, that is, discontinuities occurring through the reduction of the wave function, would be different or would be avoided if the space to which the projection is made had an other structure than the Galilean-Newtonian space.) By founding the special and general relativity theories, Einstein just had made an extremely successful revision of space-time and thus of the physical experience space, and he hoped to achieve a success when he continued his work in this direction. For this aim he developed his program of a unified geometric field theory.

To what an extent a modification of the experience space is actually possible, can only be judged by means of real measurement possibilities. What matters is not the space of everyday experience but the space of scientific (measuring) experience of physics. Thus, it is not enough to designate the abstract mathematical configuration space as the physical experience space, but this new space had to be linked with an expansion of the presupposed measurement basis of physics. Physics succeeded long ago in discovering a significance in the concepts and formalisms of classical mechanics which reach far beyond this discipline.

As described above, fundamental concepts used in the formulation of relativist field theory and quantum mechanics were defined according to the rules of classical mechanics, while physical laws underlying those theories were created using formalisms rooted in classical mechanics. From mathematics, a particular contribution was made by the Lagrange and Hamilton formalisms. There is a well-founded assumption that Hamilton's formalism can be regarded as the logical form of the dualism described here.[101]

[101] For arguments in favour of this conjecture, see H.-H. v. Borzeszkowski and R. Wahsner, *Physikalischer Dualismus und dialektischer Widerspruch.* op. cit., pp. 88-91; "Der Hamilton-Lagrange-Formalismus – die Logik der Physik?," in *Die Wirklichkeit der Physik*, op. cit., pp. 173-210.

4.1 On the Relationship between Dynamics and Measurement Theory – Pros and Contras of Constructive Scientific-philosophical Protophysics

To avoid the possible misunderstanding that the dualism of space-time and matter means a cut of the world into space-time and matter – a cut in the sense that both sides of them exist independently from each other or are to be seen as metaphysically isolated entities – we wish to make some specific remarks on the position of the principle of inertia in the Newtonian mechanics.[102] This position shows that it is not possible to define the so-called natural motion, the etalon of motion, purely for itself without taking the motions to be measured, the dynamics, into consideration. The wording of the first law clearly shows that here reference to the second law is made in the sense that one cannot define what force-free motion means without knowing how the force is defined. The relationship between the two laws also shows in the fact that the first – provided classical mechanics is elaborated completely – is a special case of the second one ($\mathfrak{F} = 0$, thus $m\mathfrak{b} = 0$) but one cannot succeed satisfactorily without its special formulation because the second law can be formulated only on the basis of the first law given by the principle of inertia. This correlation is also proved by the fact that, historically seen, the formulation of the concept *inertial system* and the development of the Newtonian dynamics were two tasks being closely linked; they had to be solved at the same time. Above all, the fact has to be taken into account that all these concepts were not created independently of the real possibilities of making experiments and observations. They are characterized by the fact that terrestrial experiments were made with respect to a system being linked with the motion of the earth, and that observations of celestial motions were at first related to the terrestrial system and later to the solar and the fixed star system. Therefore, with different degrees of accuracy, all three systems can be applied as inertial systems even today. Thus, if physics conceives the (dialectical) contradiction in the form of dualisms in such a manner that it separates the momenta existing only in a unity and distributes them to the different sides of dualism, it will always imply the interdependence of the distributed momenta in such a manner as it had been presented here with regard to the relationship between the first and second laws. This is also shown in the fact that the dynamic theory contributes to deciding what can be measured.[103]

That is the reason why the constructive scientific-philosophical program of the protophysicists cannot be implemented;[104] their intention is to construct a prephysi-

---

[102] For a detailed discussion on the protophysical arguments regarding this possible misunderstanding, see H.-H. v. Borzeszkowski and R. Wahsner, *Messung als Begründung oder Vermittlung?*, op. cit.

[103] In the same sense, M. Planck writes: "The decision about the fact whether a physical quantity can be observed in principle or whether a certain question has a physical sense, can never be taken *a priori* but only from the standpoint of a certain theory." ["Denn die Entscheidung darüber, ob eine physikalische Größe prinzipiell beobachtbar ist, oder ob eine gewisse Frage einen physikalischen Sinn hat, läßt sich niemals a priori, sondern immer erst vom Standpunkt einer bestimmten Theorie aus treffen." (M. Planck, *Das Weltbild der neuen Physik*, op. cit., pp. 37 f.)]

[104] This concept of protophysics going back to Hugo Dingler must not be mistaken for the concept given by Hans-Jürgen Treder. The latter does not determine prephysical measuring principles but

cal space-time teaching as a measurement-theoretical presupposition of physics and to establish on this basis actual physics, which is considered in this program as a discipline collecting data of measurement. On the other hand, the protophysicists satisfy an urgent philosophical demand; they try to establish proof of the fact that what was determined by Kant as the pure form of sensuousness, as the *a priori* form of intuition concerns the "measurement function" of geometry.[105] To show to which extent the demand mentioned above is satisfied by the protophysical program, we want to deal with it more comprehensively now.

The basic protophysicalists' ideas is to introduce the concept of inertial mass via the completely inelastic collision.[106] This is based on the following idea: When two totally inelastic bodies K_1 and K_2 impact upon one another with a velocity of v_1 and v_2 , they continue to move after the impact together with the velocity v. Now, the two bodies K_1 and K_2 are made to impact upon one another with different velocities and for each process of collision the quotient $\frac{v_2 - v}{v - v_1}$ is calculated. The result is that this quotient has always the same value for the two bodies K_1 and K_2, irrespectively of the velocity with which they impact upon one another. Another value occurs for two other bodies but it is again a constant value for all their velocities of impact. One concludes from this that the constant $\frac{v_2 - v}{v - v_1}$ is a quantity characterizing the bodies and interprets it as the ratio of the inertial masses m_1 and m_2 of the two bodies.

One must ask, however: Is it really the concept of mass necessary for physics? The formula $\frac{m_1}{m_2} = \frac{v_2 - v}{v - v_1}$ properly only characterizes the "totally inelastic collision" and does not indicate any possibility of measuring the mass. It only says that the ratio $\frac{v_2 - v}{v - v_1}$ is constant as to the collision of totally inelastic bodies. It is only dynamics itself showing that this constant represents the ratio of masses occurring in kinematics and later in dynamics. Without knowing dynamics, it would not be possible to see whether this is a specific quantity characterizing the totally inelastic impact or a quantity characterizing all physical motions. The necessity to integrate dynamics in the determination of physical basic quantities starts here and continues steadily. For example, to denote the ratio of the forces $\mathfrak{F}_1$ and $\mathfrak{F}_2$ acting on two bodies in an electrical

---

classical mechanics as protophysics. Classical mechanics is protophysics in the sense that in the further development of physics, one has always to come back to the original questions and basic concepts of this physical theory. (See note 89, and H.-J. Treder, H.-H. v. Borzeszkowski and R. Wahsner, *Newton-Studien*, Berlin 1978.)

[105] Thus, Lorenzen says that it is a philosophical task to understand how geometry is possible – possible as the teaching of length measurement preceding measuring physics because one has to learn measuring procedures before one gets measuring results. In this context, he points out, as mentioned above, that geometry, in contrast to arithmetics, gets its conceptions by *ideation*, and that in ideation, in contrast to *abstraction*, there are also norms of action. (Cf., P. Lorenzen, "Theorie des technischen Wissens," op. cit.; see also, H.-H. v. Borzeszkowski and R. Wahsner, "Die Natur technisch denken?," op. cit.)

[106] See, for example, P. Mittelstaedt, "Zur Protophysik der klassischen Mechanik," in *Protophysik*, op. cit.

field as the ratio of their charges e_1/e_2 is only reasonable when presupposing the laws of electrostatics. Thus, the measurement basis of physics would not be given even when one succeeded in achieving the protophysical introduction of the inertial mass. What matters additionally is to determine further basic quantities necessary for physics like, for example, electrical charge, heavy mass and so on.

Controversial opinions among the advocates of the constructive scientific-philosophical school show that for them, too, this problem in the program is not yet solved.[107] The fact that the concept of mass cannot be formulated without using dynamics, that the concept of mass thus does not – like Euclidean geometry – precede dynamics (protophysically expressed, empirical physics), caused P. Lorenzen to denote his definition of inertial mass not as protophysical but as semi-empirical.[108]

Constructive philosophy of science accepts the impossibility to establish the concept of mass without using dynamics by the fact, for example, that it considers theories of length and time measurement (geometry and kinematics) as not sufficient for establishing *all* measuring quantities because the possibility to measure the inertial mass depends on impact mechanics and to measure the charge on electrostatics. However, there is seen no problem in achieving the purely dynamic quantity *heavy mass* owing to the view that no more is necessary than to think that gravitational fields are produced by the moved inertial mass. According this one could say that neither classical nor relativistic theories of gravitation need the heavy mass. The possibility to think such a filed production cannot be meant, of course, in the sense that it is imaginable in the every-day sense or thinkable at all but only in the sense that there exists a scientific justification for it. Undoubtedly, this is given through the Newtonian or Einsteinian theories of gravitation. But this also shows again that dynamics is necessary to define the measurement quantity *heavy mass. In* a concentrated manner one could say: The answer to the question what is meant by the formulation "moved inertial masses produce gravitational fields" determines the content of the respective gravitational theory.

It is well-known that mass is introduced in the Newtonian gravitational theory in three different versions – first as resistance of inertia, as inertial mass m_I, second as passive gravitational mass m_P occurring as coupling constant by indicating how strongly a given gravitational field of the field force $\mathfrak{F} = -\text{grad}\varphi$ acts on a body; and third as the active gravitational mass m_A, that is, as the source strength of the gravitational field. These are three different quantities that can be equated only through the axiomatics of this theory. As to the first two meanings, mass occurs in the Newtonian law of motion: $m_I \mathfrak{b} = -m_P \text{grad}\varphi$. In their third meaning, they occur in the Newtonian gravitational potential: $\varphi = -\frac{m_A}{r}$. With referring expressively to his pendulum tests, Newton then postulated the equivalence of inertial and heavy masses (called above

107 For the various standpoints, see the papers in: *Protophysik*, op. cit.

108 P. Lorenzen to the authors, letter of 26 August, 1980, in *Messung als Begründung oder Vermittlung?* , op. cit., pp. 3 f. As to the determination of Euclidean geometry as measuring theory by P. Lorenzen, see, e.g., P. Lorenzen, "Geometrie als meßtheoretisches Apriori der Physik," in *Protophysik und Relativitätstheorie*, ed. by J. Pfarr, Mannheim–Vienna–Zürich 1982; *Elementargeometrie. Das Fundament der Analytischen Geometrie*, Mannheim–Vienna–Zürich 1984.

"passive gravitational mass") and puts $m_I = m_P$. Furthermore, the equivalence of active and passive gravitational masses ($m_A = Gm_P$) is established by the action-reaction law. Therefore, presupposing the Newtonian gravitational theory (that is, Newtonian laws and Newtonian force term for gravitation including the experimentally founded equivalence of m_I and m_P) heavy mass is also given by inertial mass.

If sharpening the equivalence of these three masses to the Einstein principle of equivalence with taking the principle of special relativity into account, we come to the general relativity theory. In this theory, one mass is enough again, but, to emphasize this point it should be repeated, only under the condition that general relativity is assumed to be valid.

As to our critical remark concerning to the attempt to introduce the mass protophysically, let us add a general remark: This plan is based on the assumption that the concept *mass* is already available, it ignores the difficulties linked with making them comparable and thinking them as being of the same kind. Nicholas of Cusa had to elaborate a whole philosophical system in order to solve this problem, in order to found the principle of the quantitative comparison of different qualities.

The fact that the measurement-theoretical basis of physics cannot be developed before and independently of physics is evident not only in connection with the concept of mass. It is true, the Euclidean geometry existed before the foundation of physics and all physical theories have to refer to the Euclidean geometry in the definition of measurement stipulations but pure geometry does not make any statements about the fact in what a regard it is appropriate to be used as measurement-theoretical basis of physics. For this purpose, additional efforts had to be made and these were identical with the process of the genesis of physics. Let us only remind of the fact that Archimedes had to formulate an appropriate additional postulate about the attribution of the center of heaviness of a body to a geometrical point designated on it in order to be able to use geometry as the theory in his statics, or Galilei had to take great efforts aimed at making the Euclidean geometry the theory of measuring motions. These efforts especially implied the solution of the problem how to combine Euclidean-space quantities with time quantities for this purpose.[109] The measurement basis of physics resting on the measurement quantities *space*, *time* and *mass* was only given with the foundation of classical mechanics by Newton. All subsequent physical theories became measuring physics by saying how their basic concepts were linked with the concepts of Newtonian mechanics. In this manner, and only in this manner, the Euclidean geometry precedes classical mechanics and this mechanics precedes whole physics. Only if one understand the precedence of the Euclidean geometry in this manner, one can consider it as *a priori* measurement-theoretical part of physics and determine this part as a (theoretical) means of cognition in physics. And in this sense, and only in this sense, classical mechanics is related to all the following physics as protophysics.[110]

---

109 See H.-H. v. Borzeszkowski and R. Wahsner, "Zur Beziehung von experimenteller Methode und Raumbegriff," op. cit.

110 See notes 89 and 104.

In our opinion, this term "protophysics" is justified for two reasons: On the one hand, all physical theories known so far are structured in principle in the same way as classical mechanics; thus classical mechanics is the prototype of a physical theory. On the other, the relation between any physical theory and reality can only be established by classical mechanics in the end. Thus, the term "protophysics" is being used in another sense than in the constructive philosophy of science, of course. The two terms, however, are interrelated to such an extent that both of them refer to the conditions of physics required by measurement. To our mind, however, theories of space and time are not the only conditions of physics required by measurement, but one needs a theory, in addition, saying how to combine the quantities of length and time in order to make motion measurable. Additionally, further quantities (inertial mass, active heavy mass, passive heavy mass, charge) and the determination of quantities of measurement to be formed from these quantities as well as from velocities and accelerations are necessary in order to characterize the motion. And all this is only possible within a physical theory, and classical mechanics was the first to do it. It is true, of course, when establishing a new physical theory, one has always to think about the fact what is measurable in which manner, but all theories formulated so far have shown that classical mechanics has given answers to basic questions of the measuring process that are valid now as before. As shown above, one is even obliged to realize "*quasi*-Newtonian situations" within the framework of each physical theory in order to link this theory via measurement with reality so that they can be interpreted physically.[111]

Accordingly, it is appropriate to use the term "protophysics" in order to characterize the epistemological bases of physics; however, this concept does not coincide with the concept "measurement-theoretical *a priori* presuppositions of a physical theory". The latter refer only to the space-time structure actually; they say which conditions have to be satisfied by space-time in order to make reproducible measurements possible, but taken alone, they do not yet supply positive determinations of quantities of measurement characterizing physical interaction.

According to the different concepts of protophysics, the *a priori* measurement-theoretical role of the Euclidean geometry for physics is seen different, that is, this geometry is integrated into physics differently. Here, its role is seen in the fact that in general relativity theory the space-time is *quasi*-Euclidean in infinitesimal regions and that some features of global connection are taken from the Euclidean geometry in order to make the comparison of measurements possible that are made at different places of space-time (there is no *quasi*-Euclidean background but space-time is identical with the gravitational field);[112] the constructive scientific-philosophical conception pre-

---

111 As dynamics decides upon the fact what and how can be measured, different procedures of measurement, which bring about the same results in the Newtonian theory, can in other physical theories lead to quite different results. Thus, it has to be decided individually which procedure is the appropriate one in each case.

112 Cf., H. Reichenbach, "Der gegenwärtige Stand der Relativitätsdiskussion," in *Hans Reichenbach, Gesammelte Werke in 9 Bdn.*, ed. by A. Kamlah and M. Reichenbach, vol. 3, Braunschweig–Wiesbaden 1979. – He presents the views discussed at the beginning of the 1920s in a critical way.

supposes the Euclidean geometry as the genuine space-time structure being the background for gravitational physics described by potentials.

Although the Euclidean geometry cannot be considered as the "genuine" space-time structure, it has to be considered as a measurement-conditioned presupposition for the theory of relativity and quantum theory. In particular, one has to oppose the constructive scientific-philosophical conception saying that it would not be consistent to accept the theory of general relativity as a new space-time theory, that is, a non-Euclidean theory, and to say at the same time that the Euclidean geometry precedes the theory of general relativity for measurement-theoretical reasons. When it is said that this geometry is valid in the small in the Einstein gravitational theory, it means, in more precise terms, that it is valid in the infinitesimal neighborhood of a point or, said even more precisely, in the tangent space of this point. The theory of general relativity proceeds from the fact that those spaces in which measurements are made actually, that is, the respective laboratories with their measurement devices, are Euclidean ones. And it says also that owing to gravitation, the measurement spaces distributed all over the world, namely the laboratories, cannot be united to a global Euclidean space. On the other hand, according to general relativity theory, any arbitrary composition is not admissible but only such a composition that guarantees the telecomparison of distances measured in different spaces-time points so that the possibility is given to transport rigid measures free from distortion.

Irrespective of the difference in the views discussed in the present paper, both of them need global (expanded over a finite area at least) spaces that make the telecomparison of locally achieved results of measurement possible. In the interpretation of the theory of general relativity as made by elementary particle physicists like Feynman and others, this space is equated with the global Minkowski space of the theory of special relativity. Then the gravitational potentials $g_{\mu\nu}$ like other physical potentials, too, – are purely special-relativistic tensor fields. When $g_{\mu\nu}$ is denoted as the metrics of space-time, this is – as said by this party frequently – only a geometric wording actually. Gravitation is seen in this context in such a manner that it differs from all the other interactions as little as possible. However, this does not meet the physical content of the principle of equivalence; as this principle is based on those properties of gravitation that make it different from all other physical interactions, and it is those properties that make the geometrization of gravitation possible.

According to Einstein, geometrization means that the Riemannian space with the metrics $g_{\mu\nu}$ can actually be considered as physical space-time so that the Minkowski space-time is superfluous as background. The Riemannian space can play this role as it has all the properties that are necessary for measuring and for comparison of measurement results over finite distances. Thus, the only criterion is satisfied determining whether a space-time is a real space-time in the sense of physics. Therefore, it is not simply curved spaces that are admitted but only spaces where the affinity $\Gamma^{2}_{\mu\nu}$ is formed from the derivatives of $g_{\mu\nu}$ and not introduced in addition to $g_{\mu\nu}$.

In contrast to this, the interpretation of the Einstein gravitational theory supported by the constructive philosophy of science implies to presuppose *the* global Minkowski space-time for the telecomparison of locally established space-time measurements; this is done in the sense that only a Newton-Minkowski inertial system remains as reference system by neglecting the gravitational action of (inertial) masses in a finite area. (Therefore, the followers of this conception do not hold it necessary to formulate additional principles of connection. These principles seem to be given for them.) To our mind, however, the outstanding role of the *quasi*-Euclidean, i.e., Minkowskian, geometry cannot mean to suppose the *global*Minkowski space-time as the background for all physical processes. We think that the crux of the problem is met by the constructive scientific-philosophical thesis saying that the "genuine" global space-time results if gravitation is neglected.

Undoubtedly, there exist physical situations where gravitation is so weak, as compared with other physical interactions, that it can be neglected. However, the Einstein gravitational theory is the theoretical realization of the strongest version of the principle of equivalence according to which gravitation exists everywhere for it cannot be switched off or screened in finite areas. Therefore, as mentioned above, it is possible to realize a Minkowski space in specific physical situations approximately but it basically contradicts the spirit of the general relativity theory to say that it would be possible to achieve the global Newton-Minkowski space-time by neglecting gravitation. The principle "to neglect gravitation" is not a rule for producing the genuine background. This is evident owing to the following consideration: If the above-mentioned demand is satisfied within the framework of the general relativity theory, one has to renounce of all non-gravitational matter because of the universal coupling of gravitational field and non-gravitational matter, that is, one has to put the energy-momentum tensor $T_{\mu\nu}$ of non-gravitational matter in the Einstein equations equal to zero. However, one does not get then the Minkowski space automatically but the whole manifold of vacuum equations in the Einstein theory.

It should be mentioned in this context that the further development of physics might show the necessity to introduce a global Minkowski space into physics. Such a Minkowski space could be introduced in various ways. One possibility is the following: The Einstein gravitational equations are maintained and a Newton-Minkowski background is fixed through the explicit indication of additional conditions (they can be motivated physically in different ways). Another possibility is as follows: The validity of the strong equivalence principle is contested and thus different gravitational equations are formulated right from the beginning, a procedure which may also result in postulating the Newton-Minkowski space as background. There exists a great variety of theories competing with the general relativity theory, starting from this viewpoint and being subject of lively discussions.[113] However, if a postulate determining the background is not only an unneccesary appendage of the theory, it would be necessary to prove that it helps eliminate difficulties or shortages in the Einstein theory.

Unless one presumes a modification of the Einstein theory right from the beginning where a global Newton-Minkowski space occurs in addition to the Riemann space, the latter can only be identified with the physical space or the physical space-time, that is,

113 See also H.-H. v. Borzeszkowski, U. Kasper, E. Kreisel, D.-E. Liebscher, and H.-J. Treder, *Gravitationstheorie und Äquivalenzprinzip*, ed. by H.-J. Treder, Berlin 1971; C.M. Will, *Theory and Experiment in Gravitational Physics*, Cambridge 1993.

the Riemannian space is the "genuine" space-time. However, within the framework of the constructive scientific-philosophical protophysical program, a real space is a space in which measurement rods are produced through ideation, i.e., measurement rods existing before the metrics $g_{\mu\nu}$ and independently of it. This view could be rather reasonable if it would not bring about the question how to decide whether this space can be extrapolated into space beyond our field of local experience. If this possibility of extrapolation was simply presumed, this would correspond to one of the above-mentioned additional postulates or the introduction of a Newton-Minkowski background.

As to the extrapolation, the view of constructive scientific-philosophical protophysics is that one has to start from a local Newton-Minkowski region and this has to be continued only to the extent to which the measurement of an imagined Cartesian system of coordinates are possible – that is what the limit of our region of experience is.[114] However, the assumption that this would make it possible to achieve what is necessary for physics is a postulate requiring a connection going beyond our region of immediate experience. Even most up-to-date techniques make it only possible to station clocks and measurement rods on the moon and some planets in our solar system. Many of the constructive scientific-philosophical deviations are implicit global presuppositions. However, the representatives of this theory do not recognize this fact. They argue that any additional postulates of connection are not necessary because the connection is given via the topology of Euclidean geometry (and not via another geometry, via a curved one, for example). Just by this viewpoint global properties are postulated. Each of the constructive scientific-philosophical proposals contains such a demand or is not implementable.

To avoid misunderstandings it should be pointed out that the term "global determination" does not aim at a metaphysical concept of cosmos. The term says that a scientific approach requires to relate to each other experience gained from different positions in order to objectivate them. And for this purpose, specified principles of connection have to be postulated.

In order to give the protophysical theory of science a more solid basis P. Lorenzen developed a purely form-theoretical foundation of the Euclidean geometry.[115] But that foundation can not remove the shortcomings of protophysics. It is true, that this foundation allows to construct, for example, square lattices on two basic points by applying certain basic constructions but there are no stipulations providing a "calibration connection" between geometric forms produced in the place *A* and in the place *B*. However, the measurement rods produced in these two places and also in other places have to be calibrated in a way in order to make it possible to compare the measurements made with them, or one is satisfied with a collection of *A*-, *B*-, *C*-

---

114 Even if one confines oneself to this region the following problem arises: Let us assume that it is possible to construct locally, in a small finite region, Euclidean-geometrical relations by means of the constructive scientific-philosophical norms of action. Then there remains, however, the problem to solve which rules of action could make it possible to achieve a criterion that allow to differ an Euclidean space from a space of a constant non-vanishing curvature.

115 Cf., P. Lorenzen, *Elementargeometrie*, op. cit.

and so on physics. The same problem occurs as to the measurement of time. The production of technically identical clocks in the place *A* and at the place *B* based on specified basic constructions allows respective local time measurements but no statement relating to time lapse in larger areas, in Europe, for example. For this purpose, the synchronization of identical clocks is necessary, that is, the establishment of a global connection. It is placed strong emphasis on this fact because studies of literature relating to the constructive scientific-philosophical reconstruction of modern physics clearly show that disregarding it may bring about severe misunderstandings as to the theory of relativity, above all. For example, great efforts are taken to clarify how a so-called empirically substantiated theory of relativity allows the further existence of the Euclidean-Galilean theory of length and duration measurement; this is made with the intention to make the theory of relativity acceptable from a standpoint of the constructive scientific-philosophical protophysics. However, the problem mentioned here is only an apparent one because stipulation for the production of measurement instruments and criteria for assessment of them are not subject of the respective physical theory for the physicist. In the special relativity theory, they are the same as in classical physics. Therefore, Einstein speaks about identical measurement rods and clocks. What is changed in the special relativity theory, is the principle of synchronization, the calibration of clocks. Exactly this is the meaning of the principle of special relativity.

This shows not only that the measurement theory with all the precedence of the Euclidean geometry is not completely independent of the physical theory to be constructed on its basis but it also proves that big problems are linked with the calibration of etalons separated by a space-time distance, that is, with the normalizing of measurement units. Normalizing is not primarily a technical problem because it means to establish a correction between clocks and measurement rods stationed in different places. Thus, normalizing contains a statement on the global connection. And physics shows that the global connection is not given by the conviction that one can think an Euclidean-Galilean system of coordinates from one laboratory to the other. The global connection is different from that what was assumed first. However, this new connection is not newly given by pure thinking but it can be produced owing to a new kind of experience, on the basis of scientific experience; here this especially is experience gained as to the propagation of light.

Normalizing is linked with many problems due to the fact that it is always connected with dynamics and kinematics. In this connection it is a crucial point to be mentioned against the protophysical scientific-philosophical reconstruction of physics that modern kinematics is not the antique kinematics. Of course, kinematics existed already in ancient times before the foundation of physics but it is not kinematics at all. Present-day kinematics was developed in connection with dynamics. The problem inherent in the normalizing may be hidden by the fact that it is possible to realize the calibration of remote etalons technically via given rules. But what appears to the technician as retrogradely, must not be designated as precedely. Many things the technician knows and is able to do are not resulting from his immediate experience but from physics, and physics presupposes thought determinations reaching far beyond the experience made by the technician.

If physics was taken only as a theoretical pillar of technics, one would never come to a physical theory although the technician is allowed to take physics in such a way, of course. In other terms: For an already existing physical theory, many things could be interpreted local-terrestrially or form-theoretically afterwards but one would never achieve a physical theory on this basis. To found the Newtonian theory of gravitation, for example, it was necessary to get an insight into the unity of sky and earth on the basis of natural laws. When one has such a theory, one can represent its origin, of course, in such a way as if it was based on purely terrestrial experience. But this means to deny the fact that one considers different terrestrial experience on the basis of "universalizing" or "globalizing" thought principles being the result of a unitary consideration of the terrestrial motions and motions of planets, and that these principles are part of the physical content of the theory as only they allow the formulation of laws. Thus, the classical gravitational dynamics is not developed at first and then examined whether it is suited for the explanation of celestial phenomena, but it can only be developed by synthesizing existing results on the terrestrial mechanics, on the one hand, and the motion of celestial bodies, on the other, so that artefacts produced in the experiment and not influenced natural motions only being observable are reduced to a common denominator.

Therefore, the epistemological considerations lying on the basis of any physical theory must not be seen only as thoughts that had been important heuristically in former times and are being superfluous today. The origin of physics must be explained like the origin of geometry because physical theories are not based only on the Euclidean geometry but they have their own epistemological status. They grasp an entity what cannot be grasped by any geometry – that is the motion. However, if the assumption is made that physics does not surpass geometry theoretically, it will only be necessary to explain the origin of geometry. Then, it is a consequent approach in a certain way to consider the genesis of the Euclidean geometry as the only coupling of human practice and formation of theory, although it is astonishing to see that the antique experience of production is presented as the only empirical basis of all science. This is only possible with the assumption that handicraft is the only human practice and Euclidean geometry is the only theory.

Finally, the protophysicists conceive the measurement function of the Euclidean geometry only as a handicraft standard for the production of measurement devices. They do not see the necessary physical-theoretical component. Therefore, measurement theory and dynamic theory come asunder for them. To say it more precisely: A dynamic theory does not exist for them, it is replaced by "empirical physics", a collection of results of measurement. According to this conception, the theoretical part of physics is only composed of the measurement theory which has yet to be developed by the constructive philosophy of science. This measurement theory is to justify or reject findings of physics. Hitherto, the protophysical reformulation of physics was not successfull; this is shown, as mentioned above, in the difficulties mainly linked with the attempt to introduce the mass protophysically.

Undoubtedly, the advantage of the protophysical program consists in recognizing the fact that it is not possible to found or explain physics in a purely rationalist manner but that for this purpose, it is necessary to incorporate the productive-objective (*produktiv-gegenständlich*) activity of man into the consideration. This finding is linked with the cognition that measurement and thus all measuring sciences need *quasi-apriori* presuppositions, theoretical and practical measurement devices, that physics cannot collect experience by mere sensual perception and then generalize this experience into concepts and hypotheses as the empirist believes. The rejection of a rationalistic foundation, however, tempts the constructive philosophy of science to ignore finally the necessity of thought principles (examples are those developed by Cusanus) in an empiristical manner by taking such principles only as the expression of rational arbitrariness. Or in other words, mechanics as first shape of a theoretical natural science is not conceived as synthesis of τέχνη and φύσις, i.e., of the forces of hand and the forces of nature; it is one-sided considered as a theory of the forces of hand.[116] The protophysical theory is not able to comprehend thought principles as the philosophical reflection of social activity as a whole basically because it reduces experience to handicraft experience. Only that is to be really objective that can be deduced from handicraft experience. Thus, one does not see the special character of scientific experience and the epistemological status of experimental methods of physics. Proceeding from this view, one believes that thinking and sensuousness in physics still have to be linked in a valid manner, that is, in a constructive scientific-philosophical manner.

---

116 Cf. H.-H. v. Borzeszkowski and R. Wahsner, "Die Natur technisch denken? Zur Synthese von τέχνη und φύσις in der Newtonschen Mechanik oder das Verhältnis von praktischer und theoretischer Mechanik in Newtons Physik," op. cit.

4.2 The Importance of the Galilean-Newtonian Mechanics for Physics as to Measurement theory – Exemplified by the Problem of Field Quantization

For the connection between dynamics and measurement-theoretical presuppositions as well as for the fact, that due to this connection, the dualistic distribution into space-time and dynamics has to be determined newly, an impressive example was given in the paper "Zur Frage der Meßbarkeit elektromagnetischer Feldgrössen" presented by Bohr and Rosenfeld in 1933.[117] The statements made in this work were of basic importance for the development of the quantum field theory and even today they are of great value for the further development of the field theory, especially the gravitational theory.

Bohr and Rosenfeld analyzed the connection between classical physics and quantum theory. They showed that classical physics is important for the quantum field theory in a protophysical sense because measuring bodies have to be macrophysical bodies and thus objects of classical mechanics as to essential properties. The dualism occurring in the physical determination of motion appears here in a certain manner as the connection between classical mechanics and quantum field dynamics.

Bohr and Rosenfeld demonstrated that there are two aspects under which the measuring bodies have to be considered as classical ones. First, the quantum structure of the measuring body has to be ignored, the measurement body has to be taken as classical as to its structure. Bohr and Rosenfeld write: "It is true that the measurement of fields requires the use of material charged test bodies [i.e., bodies with a quantum structure], but their unambiguous application as measuring instruments depends exactly on the extent to which we can treat their response to the fields as well as their influence as field sources on the basis of classical electrodynamics."[118] Second, measuring bodies must be uniformly charged, finely extended, rigid bodies in the sense of classical mechanics. The argument that the existence of rigid bodies contradicts the finite propagation velocity of all physical actions, that is, the theory of special relativity, has to be rejected by the argument that one must not think of rigid bodies *per se* but of an artificial product à la Galilei. This body can be an extremely complicated physical system which may differ from a rigid body in a sense of mechanics in many aspects. It has only to be constructed in such a manner that, relating to field measurements mentioned above, it closely approximates the *behavior* of a rigid body.

Bohr and Rosenfeld showed that the measuring bodies must have the two classical properties mentioned above in order to avoid contradictions between the measurement determinations and the quantum field theory. Such contradictions must, however, be avoided because their occurrence would mean that the dynamic field theory – since not resulting in measurable consequences – would have no physical meaning.

---

117 N. Bohr and L. Rosenfeld, "On the Question of the Measurability of Electromagnetic Field Quantities," in *Niels Bohr, Collected Works*, vol. 7, ed. by J. Kalckar, Amsterdam et al. 1996, pp. 123-166. [Originally published in German under the title "Zur Frage der Meßbarkeit der elektromagnetischen Feldgrößen," in *Kgl. Danske Vidensk. Selskab. Mat-fys. Meddelelser* vol. 12 (1933), No.8].

118 Ibid., p. 126.

The Bohr-Rosenfeld paper had just been written as critical remarks to works written by Heisenberg[119] and Landau and Peierls[120] who had taken the quantum structure of the measuring body into account and thus had come to contradictions mentioned above. Bohr and Rosenfeld stated: "All measurements of physical quantities, by definition, must be a matter of the application of classical concepts; and that, therefore, in field measurements any consideration of limitations on the strict applicability of classical electrodynamics would be in contradiction with the measurement concept itself."[121]

The precedeness of classical physics in the sense of the necessity to presume a classical structure of measuring bodies, however, does not mean – as repeatedly mentioned above – that one has solved for all times the question which aspects of classical physics are important or not for a new theory or where the "divide" has to be made between the measurement-theoretical part and dynamics. Here again, we come to the conclusion also resulting from the discussion about the ambiguous position of the second Newtonian law and saying that the etalon of motion cannot be gained independently of dynamics. The investigations of Bohr and Rosenfeld also show that this problem can only be solved in connection with the respective new theory. Thus, an essential difference exists between demands on a measuring body to be used for the measurement of fields of non-quantized electrodynamics and of fields of quantized electrodynamics. As to the measurement of the latter, it is true that one has to presuppose that the structure of a measuring body is described classically, but as to its displacement it has to be taken as an object of quantum mechanics and not of classical mechanics.

Bohr and Rosenfeld showed that the relations resulting from measurement-theoretical discussion also result from the quantum formalism of electrodynamics. This shows the compatibility of measurement theory and quantum dynamics proving that quantum dynamics has a physical meaning actually. The limitations imposed on the possibility to measure field quantities correspond to the uncertainties that had been calculated. Thus, effects calculated in dynamics can be measured. Assuming, on the other hand, that the quantum dynamics is ensured physically, one gets from this the proof for the correctness of the description of the measuring process as made by Bohr and Rosenfeld.

It is quite evident that the clarification of the measurement-theoretical basis of the quantum field theory is more than a subsequent justification of functioning physics but it can help develop physical theories. The discussion about the principles of measuring electromagnetic fields shows how the quantum theory of fields has to be developed. It must be such a theory that fundamental relations like the uncertainty relation arising from the measurement theory also result from the formalism of the quantum field theory. Thus, it is explained, so to speak, what it means to transfer quantum mechanics to fields.

---

119 Cf. W. Heisenberg, *The Physical Principles of the Quantum Theory*, Chicago 1930.

120 L. Landau and R. Peierls, *Zs. f. Physik*, vol. 69 (1931), p. 56 ff.

121 N. Bohr and L. Rosenfeld, "On the Question of the Measurability of Electromagnetic Field Quantities," op. cit., p. 153.

The great value of these investigations became evident, above all, in the foundation of the quantum theory of gravitational field.[122] Indeed, the arguments of Bohr and Rosenfeld were used for the gravitational field in order to get guide-lines for quantisizing the gravitational field.[123] In particular, the demands of Bohr and Rosenfeld on a physically reasonable measurement together with the principle of strong equivalence lead to the existence of a smallest length as a limitation of quantum field theory.

Since, due to the principle of equivalence, the given relations impose stronger limitations on the measuring of the gravitational field than on the measuring of electromagnetic field, one obtains a quantum theory of gravitation which is strongly different from the quantum theory of the electromagnetic field.

The principle of equivalence makes it possible to geometrize the gravitational field by identifying the potential functions describing the gravitational field with the metrical functions $g_{\mu\nu}$ characterizing the geometry of a Riemannian space-time. This so-called geometrization of gravitation can also be read reversely, namely as the physicalization of geometric quantities in the sense of their dynamization. This can be done because in general relativity theory the geometrical quantities $g_{\mu\nu}$ become physical field quantities determined by basic dynamical equations, namely the Einstein gravitational equations. As argued above, as a consequence of the principle of equivalence in connection with the quantum principle, the gravitational field cannot be quantisized for arbitrarily small distances. Therefore, on the one hand, the principle of equivalence provides the possibility to geometrisize gravitational fields and, on the other, it gives, together with the quantum principle, limits under which the developed quantum field theory loses its physical meaning, under which gravitation cannot be interpreted as a usual field. This principle makes the geometrization of gravitation (or the physical dynamization of geometry) possible and watches at the same time over that this geometrization is done only to such extent that space-time can continue to play its role as protophysically (measurement-theoretically) founded background. All this shows clearly that the dualism of measurement theory and dynamics as developed by classical mechanics and substantiated by the theory of general relativity must not be abandoned in principle unless one takes the risk to rob physics of its content.

Our considerations, performed for an example from modern physics, demonstrate the close connection between measurement-theoretical presuppositions and physical dynamics. On the one hand, just the analysis of measurement theory shows the ways towards a quantum field theory and its limits. Otherwise, this example shows also that

---

122 Cf., H.-H. v. Borzeszkowski and H.-J. Treder, "Quantum Theory and Einstein's General Relativity," *Found. Phys.*, vol. 12 (1982), pp. 1113 ff.; *The Meaning of Quantum Gravity*, Dordrecht et al. 1988.

123 See the papers, L. Rosenfeld, in *Report of the Conference on the Role of Gravitation in Physics*, Chapel Hill 1957; "Quantentheorie und Gravitation," in *Entstehung, Entwicklung und Perspektiven der Einsteinschen Gravitationstheorie*, ed. by H.-J. Treder, Berlin 1966; A. Peres and N. Rosen, "Quantum Limitations on the Measurement of Gravitational Fields," *Physical Review*, vol. 118 (1960),335 ff.; J.A. Wheeler,"On the Nature of Quantum Geometrodynamics," *Annals of Physics*, vol. 2 (1957), 604 ff.; "Geometrodynamics and the Issue of the Final State," in *Relativity, Groups and Topology*, ed. by C. DeWitt and B. DeWitt, London–Glasgow 1964; B. DeWitt, "The Quantization of Geometry," in *Gravitation*, ed. by L. Witten, NewYork–London 1962; H.-J. Treder, "Gravitonen," *Fortschritte der Physik*, vol. 11, 1963, pp. 81 ff.

the measurement determinations do not precede dynamics absolutely but dynamics also decides upon the fact what can be measured and how the measuring body has to be. In particular, the demand for the existence of a field theory the quantum effects of which can be measured, leads to presume a measuring body whose mechanism of displacement and momentum transfer are subject to the laws of quantum mechanics. For the measurement of a classical (not quantized) field theory, however, a measuring body would have been necessary which moves according to the laws of classical mechanics.

A physical theory, e.g., classical mechanics or classical electrostatics, can however only be a relatively independent measurement-theoretical presupposition for another physical dynamics, e.g., for the field theory, if there exists certain connection between presupposition and dynamics. Both must be independent principle systems of their own right but there must also be the possibility to link them. That means, under precisely given conditions, the respective dynamics must go over to the dynamics of a theory acting as the measurement-theoretical presupposition. In the so-called correspondence principles, this connection is formulated in precise physical terms.

The impression could be given that we assume every theory to be a measurement-theoretical presupposition for theories founded later as all gained knowledge is based on earlier-made findings being a relative *a priori* for it. Such a view would be contradictory to the determination of classical mechanics as protophysics given before and to the thesis about the outstanding role of the Euclidean geometry mediated by classical mechanics. However, this view is not at all implied by the standpoint represented here. This standpoint is rather a result of the fact, mentioned already several times and finding its expression in the definition of classical mechanics, and not of the Euclidean geometry, as protophysics. This fact says that dynamics co-determines on what can be measured. Thus, every new dynamics has to determine a certain part of its measurement-theoretical basis anew. As classical mechanics as protophysics remains untouched, change can only result from the relation of the new dynamics to dynamics already existing. This relation must be grasped by the above-mentioned dynamical link given by the correspondence principle.

5. Physical Measurement Quantities and the Unity of Space, Time and Matter

As physics does not perform its dualistic differentiation – its differentiation between space-time and dynamics or interaction, between particles and field, between topology and metrics, between passive and active principles – as a differentiation in a "motionless body" and an "unbodily motion", within dynamics the relationship between dynamic body, that is dynamic matter, and acceleration appears in such a manner that nothing can be said about the physical body outside of physical motion. This relationship now will be considered in more detail. To this end, the gravitational law should be used again as an example because classical mechanics achieved its greatest success by means of its theory of gravitation, and because this law is a precedent for all forces to be dealt with within the framework of Newtonian mechanics.

For the sake of demonstration, let us have a look at a gravitational system consisting of two bodies. According to the second Newtonian law $m\mathfrak{b} = \mathfrak{F}$, the equations of motion of this system are: $m_1\mathfrak{b}_1 = G\frac{M_1 \cdot M_2}{r^2}\mathfrak{r}_0$, $m_2\mathfrak{b}_2 = -G\frac{M_1 \cdot M_2}{r^2}\mathfrak{r}_0$. The first equation says that the inertial mass m_1 of the first body, multiplied by its acceleration $\mathfrak{b}_1$ is equal to the force exercised on it by the second body. This force is determined by the heavy masses M_1 and M_2 of the two bodies and their relative distance r ($\mathfrak{r}_0$ is the unit vector pointing from m_1 to m_2). Corresponding remarks apply to the second equation. The two bodies are characterized by the constants m_1, M_1, and m_2, M_2, respectively. By attaching these measurable properties to them, they are given the status of physical objects. However, the constants m and M do not designate properties of the objects for themselves but they have sense and meaning only within the framework of the theory of motion.[124] Inertia proves to be the resistance against changes in the state of motion; it differs from gravity by the fact that it is closely related to the principle of inertia and thus with the measurement-theoretical presuppositions necessary for the formulation of the theory of motion. Heavy mass, however, is a purely dynamic quantity that can be ascribed to the body only in the context of gravitational interactions. It does not make any sense to say about a single body (a body for itself) that it is heavy. Bodies can be heavy only *each to another*; this property emerges only in the interdependence. Of course, this is also valid for other interactions, e.g., for electrical interactions. It was possible to found the *science* of physical motion only on the basis of such a conception where the body with its various kinds of behavior is thought in unity with its motion.[125]

As the body shows only in motion what it is, the different kinds of mass (heavy mass or other charges and inertial mass) have a different meaning, of course, according to the different status of dynamic interaction and inertial motion. In attaching different

---

124 Therefore, instead of the term "property" one should use the term "behavior".

125 Physicists are used to say the individual body is heavy, the body has the mass m as if these were properties *per se* of a body. This speaking is acceptable as far as one knows the theory in the framework of which one is discussing. Accordingly this speaking is only a verbally incomplete version of the formalism correctly determined as to mathematics and physics.

meaning to the two kinds of mass, the dualism of space-time and matter, existing in all physics, of passive and active principles is expressed; by this inertia is mainly linked with the measurement-theoretical determination of space-time, and gravity is mainly linked with the interacting matter. In this respect the mentioned dualism also reflects the double nature of space as measurement and representation space.

The fact that the body shows only in motion what it is does not mean that the body is *dissolved* in motion. As shown as far, the physical grasping of motion requires the additional introduction of the body – of the body not as a motionless "something" or as a thing for itself (this had been just refuted). Without introducing the mass as a quantity not arising from the combination of lengths and times, neither the principle of inertia nor the second law of motion, or the term of physical force can be formulated. The body as necessary counterpole of the motion is realized in physics in the existence of constants of different types: elementary natural constants, on the one hand, mass, charge and other measurement quantities, on the other hand. The first are to be given to a physical theory both in terms of numbers and dimensions; the latter are determined only in terms of dimensions or by the stipulation saying how they are to be measured, their numerical value depends on the specific physical situation which is to be calculated in connection with the respective theory.

Hegel was probably the first to formulate explicitly that motion is the truth of the persisting, the concept of the true soul of the world, and that we are accustomed to looking upon it as a predicate or a state, although in fact it is the subject as subject.[126] He took the unity of body/matter and motion acknowledged as necessary and made it the development principle for his system. Nevertheless, in the end, he only managed to conceive the unity of body/matter and motion by regarding the former as a disappearing moment.[127] The other side to this coin is that Hegel only partially recognizes the concern of physics.

That is why he attempts *de facto* to reduce the body, the physical matter, to coagulated kinematic motion; he tries to think matter only as the unity of the motion momenta *space* and *time*, namely in the sense that the body or the mass (their notions) is produced through the superseding of the momenta "space" and "time" in unity. Hegel had the splendid idea to show that neither space for itself nor time for irself have no actuality. that both are only momenta of a whole, so that the whole is their truth. For him, this whole is the space-time unity given as *matter* and *motion*, respectively. Hegel, however, did not pose the question whether other momenta have to be supplemented to space and time in order to constitute this unity. Grasping the matter as the identity of space and time, in the so-called mechanical phenomena "momentum" and "lever", Hegel sees the passing of reality into ideality. For the sake of illustration let us quote an example given by Hegel himself. He writes: "In the *quantity of motion*, the *velocity*, which is the quantitative ratio only of space and time, also represents the *mass* and the same real effect is produced vice versa when the mass is en-

---

126 Cf., G.W.F. Hegel, *Jenaer Systementwürfe III. Naturphilosophie und Philosophie des Geistes*, ed. by R.-P- Horstmann, Hamburg 1987, pp. 16 f.; see also note 8.

127 On Hegel's conception of motion and nature, in more detail see, R. Wahsner, *Zur Kritik der Hegelschen Naturphilosophie*, op. cit., especially pp. 26-79.

larged and the velocity is deminished relatively. A brick for itself does not kill a man but it only produces this effect by the velocity achieved, that is, the man is killed by *space* and *time*."[128] Hegel's conception is to achieve – to speak in terms of modern physics– the geometrization of the matter. In doing so, the fact is neglected that for the "passing from the reality to ideality", that is, for producing a connection between theory and actuality, a bodily (*gegenständlich*) measurement is required bringing about the necessary "distribution of momenta" as mentioned above. Exactly this misunderstanding became obvious in discussing the differential calculus as a shortage in the Hegelian project of a "mathematics of nature" or a "science of measure".

Hegel determines the physical body and its relationship to motion as follows:[129] At first the mass as an inertial one has motion as resistance in itself. The real moment of differences is outside it, the motion is on it but as superseded. This is not to determine the mass as something in rest since it is the unity of the momenta *rest* and *motion* by being indifferent to both momenta. As far as it is in rest, it is in rest and does not pass to motion by itself; as far as it is in motion, it moves and does not pass to rest by itself.[130] Evidently, to be moved is identified here on this first level as the result of an external action. Progressively, from Infinite to Absolute Mechanics, the relationship of matter/body to motion is transformed, until it reaches the level of unconditionally free motion, the great mechanics of the heavens. However, insofar as Hegel only observes the curves or the trajectories, in other words, certain solutions of the physical law, he is unwittingly reducing physical motions to kinematic ones. The accelerated motion, that is, the "forced" motions actually to be described, the physical interactions, do not occur or are summarized under the limited concept of motion. Consequently, in the end, the matter is not taken into consideration as dynamic, that is, as interacting mass. The role which the second law of motion plays in Newton's concept of motion has not been understood.

Hegel's limited concept of motion is also evident when he determines the motion as the reality of space, with its opposite momenta of discontinuity and continuity, and time, with its opposite momenta of continuity and discontinuity. He says: "Motion is the infinite as unity of these opposites of time and space."[131] Although kinematic motion may be determined in such a manner, the *physical* motion as space-time unity is not yet comprehended and the existing contradiction is not only the actuality of time

---

128 "In der *Größe der Bewegung* vertritt ebenso die *Geschwindigkeit*, welche das quantitative Verhältnis nur von Raum und Zeit ist, die *Masse*, und umgekehrt kommt dieselbe reelle Wirkung hervor, wenn die Masse vermehrt und jene verhältnismäßig vermindert wird. Ein Ziegelstein für sich erschlägt einen Menschen nicht, sondern bringt diese Wirkung nur durch die erlangte Geschwindigkeit hervor, d.i. der Mensch wird durch *Raum* und *Zeit* totgeschlagen." (G.W.F. Hegel, *Enzyklopädie der philosophische Wissenschaften. Zweiter Teil. Die Naturphilosophie*, op. cit., p. 57 f. (§ 261 R)). – For the sake of better understanding it should be addedthat, for Hegel, the unity is an ideality and thus a higher thought than reality.

129 Cf. ibid., pp. 64-108 (§§ 263-271).

130 Cf. ibid., p. 65 (§ 264 A).

131 "Die Bewegung ist das Unendliche als Einheit dieser Entgegengesetzten der Zeit und des Raumes." (G.W.F. Hegel, *Vorlesungen über die Geschichte der Philosophie I*, op. cit., p. 310.)

and space.[132] When Hegel writes: "motion is connection"[133] one can agree to it when one adds the remark that for a consequent dialectical conception, there is no connection (and thus motion) without objects which are connected and which produce the connection by their behavior.[134] In other words: The "connection" cannot be thought if it is abstracted from objects producing it – or one presumes the connection by an rationalistic setting. Indeed, Hegel grasps the motion produced by the conflict of opposite momenta and the thus arising connection but he thinks these momenta as vanishing ones and thus makes the connection, the motion, an autonomous being that comprehends the momenta which produced it only as discarded shapes.

The determination of motion as the pure space-time connection – given by Hegel – is, in some sense, justified with regard to physics as far as only the motion etalon is taken into consideration. But a measurement rod is not identical with what can be measured by its help, of course. The motion etalon is not yet the physical motion. But independently of it it cannot be understood *per se* inasmuch as it had been formulated with regard to dynamics from the beginning. If one neglects this regard one has not comprehended the etalon, either.

The Hegelian view to think the motion is based on the idea to identify empirical sciences and the philosophical system of empirism; this, in the final analysis, implies not to make a difference between epistemological status of philosophy and empirical mathematized science.[135] No doubt, philosophically thought, the moving world is the unity of space and time. It is it because the philosophical thought thinks the differentiated momenta in their concrete unity where, of course, the ability to act is included. However, the physical thinking implies – since it must divide the world into space and time and the matter moving in the space-time – that physical space and physical time (*per se*) do not yet contain the ability to act. Even this was the precondition to define the physical space and the physical time, more precisely, the etalon of motion. Therefore, the ability to act must be introduced additionally as the third quantity in form of the mass so that it is possible to grasp real motion. It does not result from the unity of physical space and physical time, unless the missing fact is put into the unity as such, into the notion of unity, of connection. Thus, it is quite understandable why Feuerbach designated the Hegelian philosophy as rational mystics.[136] The physical ability to act, the mass, does not arise by the unity of physical space and physical time but space and time can be brought to a unity only through the mass. Possibly, the Hegelian concept of motion as unity of space and time aimed at this. Such an assump-

---

132 See the opinion of Hegel to this point, ibid., pp. 307, 309 f.

133 "Bewegung ist Zusammenhang." (Ibid., p. 312.)

134 This is also required for a dialectical conception grasping the motion itself as the subject.

135 According to Hegel, the position of the thought to objectivity in empirism is not identical with that in true philosophy; that position is only a moment of the position of the thought to the objectivity in dialectical philosophy. But this difference is not the point here. What matters is that Hegel measures the empirical sciences by the philosophy standard, namely, by identifying they with empirism. As to the difference between the philosophical and physical concepts of space, see also the literature cited in note 25.

136 Cf., L. Feuerbach, "Zur Kritik der Hegelschen Philosophie," in *Werke*, ed. by W. Schuffenhauer, vol. 9, Berlin 1970. p. 53.

tion is allowed. But the fact that this concepts of space and time are not determined measurement-theoretically but philosophically speaks against this interpretation. It is the determination by such a philosophy that only partly reflects the specific requirements of a measuring and calculating science although it speaks about their concepts.[137] When the Hegelian space and Hegelian time are brought to a unity, it will not result in anything.

This shortage is mainly due to the fact that Hegel did not recognize that physics does not make its assertions and statements not immediately about the real objects but about quantities and their relations.[138] This is valid in the same sense as one says that the use-value as use-value lays beyond the horizon of political economics.[139] The well-known differentiation between use-value and value made in economics is of general epistemological relevance as it concerns the differentiation between a concrete object, a concrete entity, and a understanding entity *(Verstandesgegenstand)*.[140] The reason for this differentiation results from the requirements of measurement. Real natural objects can only be determined in their mutual opposition, in their connections; they are distinguished but not separated from each other. However, by measurement and comparison, only objects can be determined that are separated from each other. Therefore, if one wants to measure the motion, one has to divide the momenta of motion and therefore to unfold actuality into space and time and into the matter moving in space-time.

This cannot result in the assumption that physics is metaphysical because the physical quantities are not the genuine objects of cognition. When the presumption is made that actuality *per se* is so constituted as physics thinks the world, as physics must think it, this really would be metaphysics. But this presumption is not justified because the quantities of physics are thought entities constructed on the basis of real equalities that are used to recognize the genuine object of cognition, the actuality, the concrete objects in their connections.

Of course, the recognition of measurement requirements, of the necessity to separate the momenta of motion does not yet provide an exact determination of concepts. One cannot predetermine what is to be defined as space and time or as matter (or mass) in a physical theory; these determinations depend on the measurement-theoretical possibilities that have to be found, as mentioned at the beginning, by hard work.

137 To the extent that the physical method coincides with the analytical method, Hegel reflects the specific concerns of a measuring and calculating science. To the extent that they differ (and it has been demonstrated that this happens), he does not reflect them. (Cf., e.g., R. Wahsner, *Zur Kritik der Hegelschen Naturphilosophie*, op. cit., especially pp. 57-60; "'Der Natur ist die Äußerlichkeit eigentümlich...'. Zur epistemologischen Position der Naturphilosophie Hegels," in *Hegels enzyklopädisches System,* ed. by H.-C. Lucas und B. Tuschling, Stuttgart-Bad Cannstatt (in print).)

138 Cf. R. Wahsner, *Naturwissenschaft*, Bibliothek dialektischer Grundbegriffe, ed. by A. Hüllinghorst, Bielefeld 1998 and the literature cited therein; H.-H. v. Borzeszkowski and R. Wahsner, *Die Wirklichkeit der Physik*, op. cit.

139 Cf., K. Marx, *Zur Kritik der politischen Ökonomie*, in *Karl Marx und Friedrich Engels, Werke*, Berlin 1956-1990, vol. 13, p. 16.

140 The notion *understanding entity (Verstandesgegenstand)* is to be used here, following the terminology of Hegel, but not completely in line with him; it is used for an abstract or ideal thing, a thing of thought in contrast to both, a sensuous-concrete and a philosophical-concrete noon.

The change in measurement-theoretical possibilities modifies also the separation in space and time, on the one hand, and into matter, on the other. In no case, however, the physical concepts of space, time, matter (or mass) are identical with the philosophical notions of space-time and matter. The determination of the fundamentals of physics under the heading of the theory of measurement implies their determinations such that one may establish on their basis a dynamics (containing the laws of motion) whose statements can be experimentally tested. Therefore, the physical concepts of space and time and matter are no pure rationalistic settings; neither they are something immediately perceptible. They are measurement-theoretical thought entities. As such they are neither given naturally nor immediately existing in human idea. It took centuries of philosophical and technical effort before mechanics succeeded for the first time in history in measuring motion. This called for a process which began with the ancient atomists and continued with Aristotle, Archimedes, Nicholas of Cusa, Kepler and Galileo forward to Newton.

The quantities, the measurement-theoretically determined thought entities, must be produced by means of comparative work. This results from the fact that a comparison can only be made under a certain aspect. Helmholtz, for example, writes that "quantities to be declared as being equal are equal to each other only *in a certain relation*". He continues: "For deciding about equality, it is necessary to know the method according to which the comparison is to be made. ... At the same time we see that the inequality is the more general case and the equality is specific exception."[141] Different things can only be compared to each other, can only be measured one on the other, if one has found a point of view under which they are equal. This demand is especially evident when one looks at sciences that get ready to become calculating and measuring sciences like biology and economics. The great difficulty is to find this quality. To achieve this, the different objects must be regarded as something what is different from that what they are as things, products, use-values, as concrete objects.

Since physics needs quantities, and since those quantities represent in each case only one quality, one equality, one must introduce all qualities necessary for conceiving a connection as separate entities. Accordingly the different momenta must be divided. Otherwise, these qualities must not diverge. Therefore, it is also necessary to formulate their connection, namely by formulating the fundamental laws of a physical theory. This relationship of dependence and independence is shown, for example, in the fact that the quantity of mass cannot be deduced from the quantities of space and time but the quantities of space and time can only brought to an unity via the quantity of mass.

Unconsciously, Hegel himself indicated the profound reason for the epistemological status of quantities and its consequences. The reason is that one side of the contradiction is firmly established, that one of the opposite predicates is put as

141 Es "sind die als gleich erklärten Größen nur in *gewisser Beziehung* einander gleich. Es gehört, um über die Gleichheit zu entscheiden, die Kenntniß der Methode dazu, nach der die Vergleichung ausgeführt werden soll. ... Zugleich bemerken wir, daß die Ungleichheit der allgemeinere Fall ist und die Gleichheit einen besonderen Ausnahmefall bildet." (H. v. Helmholtz, *Einleitung zu den Vorlesungen über theoretische Physik*, Leipzig 1903, p. 26.)

essence.[142] The respective quality of a quantity, the aspect contained in it under which certain concretely different things are equal, is the firmly established essence. With regard to a theory, a physical theory, the side of the dualism grasped by passive principles and by measurement-theoretical presuppositions is that side of two opposite predicates which is setting as essence. However, Hegel did not recognize the productive use of this mode of thinking within the framework of a measuring and calculating science. He criticizes it as external dialectics. Indeed, it is external to separate the opposite determinations of the contradiction in essence and (unessential) behavior of this essence.[143] In principle, this rupture can be put at any place. However, when the above-mentioned mode of thinking is used measurement-theoretically, this arbitrariness will be eliminated because the problem to solve is by which divide the measurement will be possible.[144] And then, the counterpole of the essence is not unessential behavior any more. It is behavior described by dynamics. Thus, it is that what makes physics. The essence shrinked together to the (although very important) presupposition.

As one side of the contradiction is set as the essence, it must be presupposed, elected on the basis of mutual consent. For this reason, physics cannot else but using thought determinations without their further study.[145] Therefore, the statement is valid: "No (special) science proves its own principles and submits them to discussion but takes them as certain *per se* ... ."[146] In the end, philosophy is necessary to found these principles. This foundation requires to supersede the setting of one of the opposite determinations as the essence.

For this purpose, the exchange processes, the "alternative actions", lying on the basis of the forming of quantities, must be shown and studied philosophically (that is,

---

142 Cf., G.W.F. Hegel, *Vorlesungen über die Geschichte der Philosophie I*, op. cit., pp. 301-305.

143 This may be the preliminary stage for both internal dialectics and sophistics; Hegel showed both trends of development.

144 More generally, one can say: Where the cut can be made, depends on the fact what is essential in the respective situation. Thus, the justification of the cut is determined locally. As an empirical mathematized science needs a cut for reason of measurement and the formation of quantities, as such a science is always based on cut, it can never be world wisdom or philosophy, and cosmology can never be usual physics. On the problem of cosmology see, H.-H. v. Borzeszkowski and R. Wahsner, "Kosmologie – Physik oder Metaphysik?," *Dt. Zs. f. Philosophie*, vol. 26 (1978), pp. 242-246; "Physikalische Erfahrung und die Welt als Ganzes," in *Die Wirklichkeit der Physik*, op. cit., pp. 287-339.

145 Cf., G.W.F. Hegel, *Vorlesungen über die Geschichte der Philosophie III*, in *Werke*, op. cit., vol. 20, p. 364.

146 This profound opinion was formulated by Proklos in his comments on Euclid's *Elements*: "As we state that this science, geometry, is based on presumtions and proves the conclusions proceeding from specific principles, ... the author of a book on elementary geometry must, teach the *principles* of the science and the *conclusions* from the principles separately; he need not give account of the principles but of the respective conclusions. No science proves its own principles and presents them to discussion but holds them to be certain *per se*; they are more clear to it than derivations; it recognizes the first in their own light, but derivations via principles ... . If anyone lumps the principles and derivations together, he creates confusion in the whole science and mixes what has nothing to do with each other. The *principle* and the respective *derivation* are separate things originally." (Proclus Diadochus, *Kommentar zum ersten Buch von Euklids "Elementen,"* ed. by M. Steck, Halle/Saale 1945, p. 218 – German translation.)

the origin of the presupposed thought determinations must be proved). This is the only way to learn which equalities, i.e., equalities of which concrete objects and motions, the understanding entities determined measurement-theoretically represent; this enables one to lead the characteristic features of thought entities back to relationships of concrete objects, i.e., to supersede the distribution of momenta. [147]

If one thinks motion as a pure space-time unity it is not surprising that one receives physics adequately only to such extent as the ignoring of mass is still acceptable, as the question is the force-free motion. That, in the end, is the reason why the epistemological importance of second derivatives occurring in physics (characterizing accelerated, dynamic motions) was recognized neither by Kant nor by Hegel. [148] They place the forces wrongly; they use them in order to overcome the gap between rest and motion (a gap that, as mentioned above, does not exist in physics) and nothing is left for the physical motion that is to be explained actually. This even is valid for the dialectical conception of Hegel. [149] The misconceiveness of the self-existence of nature, of the objectiveness (*Gegenständlichkeit*) of motion and thus the role of second derivations is connected with the neglecting of the importance of the constants.

The self-existence of the nature does not mean the independence of an individual object. Such an object cannot be an independent entity in its essence as it does not exist in an isolated manner in its essence; its existence and constitution depend on other natural objects, it was created by them and it is in connection with them. The natural objects act mutually on each other. It is true these mutual actions, this conflict

---

[147] We should here remind of the fact that in the Hegelian system externality is a characterisic feature of nature making differences disintegrate and them appear as indifferent existences. ["Der Natur ist gerade die Äußerlichkeit eigentümlich, die Unterschiede auseinanderfallen und sie als gleichgültige Existenzen auftreten zu lassen ... ." (G.W.F. Hegel, *Enzyklopädie der philosophische Wissenschaften. Zweiter Teil. Die Naturphilosophie*, op. cit., p. 31 (§ 249 R).)] Therefore, it is consequent that Hegel formulates the following thesis being essential for his conception of contradiction: "The powerful, strong, spiritual is what can endure the opposite in itself, the mind can endure the strongest contradiction – the weak physical not, it disappears as soon as another comes to it. ["Das Mächtige, Kräftige, Geistige ist das, was den Gegensatz in sich ertragen kann; der Geist kann den höchsten Widerspruch ertragen, – das schwache Körperliche nicht, es vergeht, sobald ein Anderes an dasselbe kommt." (G.W.F. Hegel, *Vorlesungen über die Geschichte der Philosophie II*, in *Werke*, op. cit., vol. 19, p. 79.)]

[148] See H.-H. v. Borzeszkowski and R. Wahsner, Newton und Voltaire, op. cit., especially pp. 45-49; "Introduction" to *Voltaire, Elemente der Philosophie Newtons/ Verteidigung des Newtonianismus/ Die Metaphysik des Neuton*, op. cit., especially pp. 36-41; Voltaire's Newtonianism. A Bridge from English Empiricism to Cartesian Rationalism and Its Implications for the Concept of Mechanics in German Idealism," Preprint 147 of the Max Planck Institute for the History of Science, Berlin 2000, pp. 8-12, 25 [also: Appendix to the present volume, especially pp. 89-95, 105-107]; "Christian Wolff's Mechanical Philosophy: A Comparison with Isaac Newton's Mechanics," Preprint 147 of the Max Planck Institute for the History of Science, Berlin 2000, pp. 8-10 [also: Appendix to the present volume, especially pp. 124-126]; R. Wahsner, *Das Aktive und das Passive*, op. cit., especially pp. 114-130; *Zur Kritik der Hegelschen Naturphilosophie*, op. cit, pp. 114, 123, (39, 48-51, 66 f.); "Die Kantsche Synthese von Leibniz und Newton und deren Konsequenzen für den Mechanik-Begriff des deutschen Idealismus," in: *Tagungsband des 9. Internationalen Kant-Kongresses, Berlin, 26. 3.-31. 3. 2000* [in print] [also: Preprint 111 of the Max Planck Institute for the History of Science, Berlin 1999]; "'Der Gedanke kann nicht richtiger bestimmt werden, als Newton ihn gegeben hat'," op. cit., especially pp. 295-300.

[149] Cf., R. Wahsner, *Zur Kritik der Hegelschen Naturphilosophie*, op. cit.

of natural objects may provoke that specific objects become independent from each other relatively and locally (not all things are immediately dependent from each other).[150] This mutual actions may also provoke, however, that an immediate dependence is created between them in such a manner that all of them have the other as their condition for real existence. These objects form a collective unit, a system. And only a collective unit or system of objects can produce itself; the system exists only because and through the fact that it produces itself. In this sense, in changing itself it preserves itself.[151] The system is contradictory. Independent of its conflict with other objects of the same collective unit, the object is not contradictory. Not objects in the thought separation of their behavior are contradictory but their behavior in a collective unit. Something is contradictory by being bodily (*gegenständlich*) behavior in a concrete correlation.[152] This concept of contradiction, arising from the thought of the independence of nature, is not only linked with the thesis of the historical character of nature inseparably but it is the only one which is possible to found it. The thought to explicate the history of nature, can only be performed when the fact is recognized that a real natural object can only be produced by another natural object. Only a system, a collective unit of natural objects, can produce itself; and it exists only because it produces itself and by doing so. Of course, this self-producing absolutely exists only for that collective unit which constitutes nature as a whole. With this view in mind, one should appropriate critically the Hegelian thesis saying that all things *per se* are contradictory.[153]

As far as Hegel does not think the bodily motion, he also does not think the actual contradiction. He replaces the objective (*das Gegenständliche*) by his concept of unity. Thus, it is consistent for him to consider the physical body as a body being contradictory *per se* as it is the unity of rest and motion and the motion is the unity of space and time. Hegel's conception of contradiction results in his inability to understand physics as a science that presupposes the ability of natural objects to act as a essential momentum, to understand physics as a theory basing on a bodily and, of course, theoretical measurement and thus essentially differing from mathematics.

Hegel's conception of the relationship between body and motion represents the typically modern-age (*neuzeitlich*) standpoint *as far as* it determines the substance as motion and not, as their ancient predecessors, as movable object.[154] It was only this

---

150 There is no object (*Gegenstand*) that had not any object or any objects as preconditions for its existence and that is not the condition for the existence of other objects. But a *specified* object can be independent of *specified* other objects, of course.

151 See the literature cited in the essays of the following note.

152 Cf., P. Ruben, *Dialektik und Arbeit der Philosophie*, Cologne:1978, pp. 52-85; R. Wahsner, "Was ist die Natur?," *Wissenschaft und Fortschritt*, vol. 26 (1976), 344-349; "It is Not Singularity that Governs the Nature of Things. The Principle of Isolated Individual and its Negation by Marx in his Doctoral Thesis 'Difference Between the Democritean and Epicurean Philosophy of Nature'," *Poznan Studies in the Philosophy of Science and the Humanities*, vol. 60, ed. by R. Panasiuk and L. Nowak, Amsterdam-Atlanta 1998, pp. 99-111.

153 "Alle Dinge sind an sich selbst widersprechend ... ." (G.W.F. Hegel, *Wissenschaft der Logik. Zweiter Teil*, op. cit., p. 74.)

154 Hegel determines the motion as substance as follows: The living substance is the being which in truth is the subject or ... which in truth is real, only insofar it is the motion of setting itself or the

standpoint which made it possible to develop natural sciences, physics in particular. But this standpoint does not yet suffice to comprehend it philosophically. In physics, the bodily, objective motion as substance was practiced, and in philosophy, the pure motion, the independent, unbodily motion was thought.

---

mediating of becoming different itself with itself. As subject, the substance is the pure, simple negativity. ... The truth is the whole. The whole, however, is merely the essence reaching its completeness through the process of its own development." ["Die lebendige Substanz ist ferner das Sein, welches in Wahrheit *Subjekt* oder, ... welches in Wahrheit wirklich ist, nur insofern sie die Bewegung des Sichselbstsetzens oder die Vermittlung des Sichanderswerden mit sich selbst ist. Sie ist als Subjekt die reine *einfache Negativität*, ... Das Wahre ist das Ganze. Das Ganze aber ist nur das durch seine Entwicklung sich vollendende Wesen." (G.W.F. Hegel, *Phänomenologie des Geistes*, in *Werke*, op. cit., vol. 3, pp. 23 f.)]

6. The Measurement Quantity Time – Direction of Time and Irreversibility

The insufficient consideration of the specific features of physical notions inherent to the space-time-unity found in Hegel is also typical of certain discussions between natural scientists on the physical time conception.

Due to the epistemological character of physical measurement quantities as thought entities, time is not introduced into physics as concrete time. As concrete time, time is directed, to use other words – it is the unity of the past, present and future. Time as a quantity of measurement does not have this meaning.[155] A certain explanation of this fact was already given by d'Alembert. He writes: "One cannot compare with each other two things of such a completely different nature as space and time: One can, however, compare the relation of the parts of time to each other with the relation of parts of the traversed path to each other ... Without knowing time in itself and without having an exact measure of it, we cannot any clearer imagine the mutual relation of its parts than we can the mutual relation of the parts of a straight line. Now the relation between the ratios of the parts of such a line and the ratios of the parts of the path covered by a body somehow moving can always be expressed by an equation. The equation will not express the relation of times to line segments but, so to speak, the relation between the ratio of the parts of time to their unit and the ratio of the parts of the covered path to the unit valid for them."[156]

Thus, it is one of the peculiarities of physics to deal with the time (like space and body) from the viewpoint of measurability (i.e., "of the mutual relations of its parts") and not to determine the time at all. Therefore, the question what does physics say about *the* time (or *the* space or *the* body) is a question which physics has not to answer primarily. The time determined in physics for reasons of measurement is necessarily reversible because it is determined in connection with the reversible, i.e., time-inversion-invariant dynamic basic equations of this science (like the Newtonian equations of motion, the Maxwell equations and the Schrödinger equation).[157] Therefore, the character of physical concepts is misconceived when one nowadays repeats again the old argument that physics should change its concept of time because the "genuine" time is irreversible or the irreversible time is so real as the reversible one.

---

155 Cf., H.-H. v. Borzeszkowski and R. Wahsner, "Did the Nonlinear Irreversible Thermodynamics Revolutionize the Classical Time Conception of Physics?," *Foundations of Physics*, vol. 14 (1984), 653-670.

156 "On ne peut comparer ensemble deux choses d'une nature différente, telles que l'espace et le tems; mais on peut comparer le rapport des parties du tems avec alui des parties de l'espace parcouru. Le tems sa nature coule uniformément, et la Mechanique suppose cette uniformité. Du reste, sans connoître le tems en liu-même et sans en avoir de mesure précise, nous ne pouvons représenter plus clairement le rapport de ses parties, que par celui des portions d'une ligne droite indéfinie. Or l'analogie qu'il y a entre le rapport des parties d'une telle ligne, et celui des parties de l'espace parcouru par un corps qui se meut d'une maniére queiconque, peut toujours être exprimée par une equation: ... l'equation de cette courbe exprimera, non le rapport des tems aux espaces, mais, si on peut parler ainsi, le rapport du rapport que les parties de tems ont à leur unite, à celui que les parties de l'espace parcouru out à la leur," (J.B. d'Alembert, *Traité de Dynamique*, op. cit., p. VIII.)

157 In this section, the totality of all dynamic theories whose basic equation are invariant against the time inversion are to be designated as "dynamic physics".

Notwithstanding this fact, it is, however, justified, within the framework of physics, to ask for a directed, "more concrete" time. This results from the second law of thermodynamics saying that there exists a function S, the so-called entropy, attributable to all macroscopic systems (composed of many particles whose motion is described by the basic laws of dynamic physics), and that function is growing with time (defined in concepts of measurement theory) monotonously: $dS/dt \geq 0$. This function determines a direction of time, namely the direction of time for which S increases.

Various authors with single aspects of irreversible processes are, due to the non-clarified relation between dynamics and thermodynamics, induced to believe that irreversible thermodynamics is a completely new type of physics that, contrary to dynamic physics, could, for instance, be used to describe development. This opinion, contrasting with the comprehension of classics like Maxwell, Clausius, Boltzmann, Gibbs, Helmholtz, and Planck, requires to study the character of the expansion of physics caused by thermodynamics for understanding the physical grasping of motion.

Even the opinion saying that a time was introduced by thermodynamics, that grasps the development from past to future, is linked with problems. According to Boltzmann, the statistical weight of a state of a quasi- isolated system is the greater, the closer it is to the final state of the maximum entropy of the system. Thus, the variety of possible prehistories grows with the approach of the system to the final state. Accordingly, the direction of time defined by the second law is characterized by the fact that the future becomes more determined and the past becomes more undetermined;[158] this is in contrast to the differentiation between past and future within the framework of a concrete concept of time.

Léon Rosenfeld, who worked on basic problems of irreversible thermodynamics together with Prigogine and other authors, rejected this exaggerated interpretation of irreversible thermodynamics by the argument that the second law cannot found a directedness of time valid for physics as a whole and that, in addition, the rupture between dynamic physics and thermodynamics is not as deep as pretended so often.[159] Rosenfeld remarked that it is logically acceptable to relate the direction of time to a physical standard phenomenon like the transport of heat (this determination is similar to the definition of reversible physical time which was determined trough the inertial standard motion, see Section 4), but this approach does not satisfy because it does not show the origin of the differentiation between past and future.[160] If one wants to elucidate this difference physically, one will have, according to Rosenfeld, to use the principle of causality being fundamental for physics. Therefore, the direction of time occurring in the second law of thermodynamics should be proved as a special consequence of dynamic physics. Even more so because the rupture between dynamic physics and thermodynamics is not so deep as pretended so often since features of a

---

[158] Cf., H.-J. Treder, "Boltzmanns Kosmogonie und die räumliche und zeitliche Unendlichkeit der Welt," *Wiss. Zs. Humboldt-Univ. zu Berlin. Math.-Nat. R.*, vol. XXVI (1977), 13 ff.

[159] See, for example, L. Rosenfeld, "Statistical Causality in Atomic Theory. A General Introduction to Irreversibility," in *Selected Papers of Léon Rosenfeld*, op. cit.; C. George, I. Prigogine, and L. Rosenfeld, *Det Kgl. Danske Vidensk. Selskab., Mat-fys. Medd.*, vol. 38 (1972), 1 ff.

[160] See L. Rosenfeld, "Statistical Causality in Atomic Theory," op. cit., p. 551.

direction of time can already be seen on the dynamic level. This is evident when one pays attention not only to the basic dynamic laws under the time inversion but also to their solutions. The latter can be time-directed. Examples are the retarded potentials that can be calculated from the Maxwell equations under certain boundary conditions and that show the correlation between the occurrence of a directed time and the validity of the principle of causality. On the one hand, these potentials define a direction of time because a light signal emitted at the time t_1 can only received at the time $t_2 > t_1$, so that $t_2 - t_1 = r/c > 0$ (r is the distance between the places of emission and absorption, c is the speed of light). On the other, they determine the light cones of the theory of special relativity and thus the possible causal nexus of events.

But if there is no fundamental rupture between dynamic physics and thermodynamics as time-directed solutions can be deduced from dynamic basic equations through adding specific supplementary conditions, it should be possible to deduce the second law of thermodynamics from the laws of dynamic physics also by adding specific supplementary conditions. Accordingly, the crux of characterizing thermodynamics is to find an answer to the question whether he transition from reversible dynamics to irreversible thermodynamics can actually be deduced from auxiliary or additional condition, and second which status has to be attached to these conditions.

It is the classical work *Weitere Untersuchungen über das thermodynamische Gleichgewicht zwischen Gasmolekülen*[161] written by Boltzmann, above all, that shows the relation between dynamic physics and thermodynamics. In this paper, Boltzmann examined a gas composed of many particles and assumed that the time change of the velocity distribution $f(r,v,t)$, i.e., $\partial f/\partial t$,[162] can be divided into two parts; the first part relates to that motion, which is force-free in the sense of mechanics between the impacts of the particles and the second relates to the (elastic) collisions of the particles: $\frac{\partial f}{\partial t} = \left(\frac{\partial f}{\partial t}\right)_F + \left(\frac{\partial f}{\partial t}\right)_S$. The term resulting from the free motion can be calculated on the basis of the equations of motion in classical mechanics, the calculation of the impact term $\left(\frac{\partial f}{\partial t}\right)_S$ needs additional statistical assumptions relating to the frequency of impacts (*Stoßzahlansatz*). This calculation led Boltzmann to a kinetic differential equation for f saying that the function $H = \int dvf \log f$ diminishes monotonously: $dH/dt \leq 0$. As H defines the negative entropy S, this is what the second law of thermodynamics says.[163]

---

161 Cf., L. Boltzmann, *Wiener Berichte*, vol. 66 (1872), 275 ff.

162 Here f denotes the number of particles that have the place r and the velocity v at a given time point t.

163 The assumption made by Boltzmann in this context saying that the hypotheses used by Clausius and Maxwell suffice to deduce the increase of entropy with growing time brought about a controversial discussion. Loschmidt (1876) and Zermelo (1896), above all, opposed this thesis with statements known in literature as *Umkehreinwand* or *Wiederkehreinwand*. The situation bacame clear after the work "Begriffliche Grundlagen der statistischen Auffassung in der Mechanik" by P. and T. Ehrenfest who showed that the irreversibility of the motion in the direction of an increase in entropy is a statement on the most probable value of the change in entropy and that the real change in entropy varies round this most probable value according to the much smaller but not zero probability

The Boltzmann derivation clearly shows that and how the second law results from the dynamic equations of a system composed of many particles under specific statistical additional conditions. The direction of time, that is, the preference of the direction of time for which H decreases and S increases, results from $\left(\frac{\partial f}{\partial t}\right)_S$, that is, the term into which those (statistical) conditions are introduced. Thus, the second law amalgamates the basic equations of dynamic physics with certain additional conditions to a new law. The answer to the question, how general this law is, as compared with basic dynamic equations, essentially depends on the degree of generality of those conditions.

It is important to state that Boltzmann's *Stoßzahlansatz* concerns those initial values that have to be presumed if the system is to be described on the basis of equations of dynamic physics but it cannot be reduced to such statements. For, statistical theorems are valid only for system that contain an infinite number of particles. A particle gas described in classical mechanics, however, is always composed of a finite number of particles so that only a finite number of dynamic initial values can be imposed here. The transition from classical mechanics (or more generally: from dynamic physics) to thermodynamics is thus a transition from the finite to the infinite, a transition that is possible only with adding new principles. This finds also its expression in the fact that thermodynamics introduces the new universal constant k (Boltzmann constant) in addition to the universal constants of dynamic physics (Planck constant h, vacuum velocity of light c, Newtonian gravitational constant G); it connects the thermodynamics and the statistical mechanics of Boltzmann and Gibbs.

Thus, the direction of time results always from additional conditions: on the purely dynamic level, from boundary and initial conditions that make it possible to calculate specific time-directed solutions and, on the thermodynamic level, from the statistical postulates. One could say that, in this manner, a historical component is being introduced into physics. However, this is not made through the transformation of the basic dynamic laws of physics. Those laws do not contain this component, but they are compatible with it, due to the possible choice of additional conditions. And these additional conditions are given by the Boltzmann-Planck principle of elementary disorder.

This is also the basis of possibilities arising from the non-linear thermodynamics of irreversible processes. But this thermodynamics is not a new evolutionary physics. Even not if *development* is not taken in the philosophical sense but, for example, in the sense of the term of biological evolution. Physical laws provide conditions necessary for the process of biological evolution[164] but these processes are not "explained" by

also of other values. (P. and T. Ehrenfest, in *Collected Scientific Papers*, Amsterdam 1959. p. 258.) Paul and Tatyana Ehrenfest said that this statement was substantiated by studies made by M. v. Smoluchowski and A. Einstein relating to the Brown motion. Their exact conceptual analysis proved that the objections of Loschmidt and Zermelo could not prove any internal contradiction in the fundaments of statistical thermodynamics. The Ehrenfests by this clarified the character of statistical laws basically, and many subsequent works are based on their work. (Cf., e.g., H.-H. v. Borzeszkowski and R. Wahsner, "Ein streitbarer Physiker im Zentrum der modernen Physik. Zum 100. Geburtstag von Paul Ehrenfest," *Wissenschaft und Fortschritt*, vol. 30 (1980), 59-62.)

164 See, e.g., W. Ebeling and R. Feistel, *Physik der Selbstorganisation und Evolution*, Berlin 1982.

those laws but by assumptions that have to be made additionally. Progress made in non-linear irreversible thermodynamics far removed from equilibrium is accompanied by a dissolution of universal laws into a variety of (very important) cases, and one must add, of course, that this is a quite natural procedure in applying basic physical laws.

Of course, the further development of the Boltzmann-Gibbs classical thermodynamics to linear and non-linear thermodynamics of irreversible processes by I.L.F. Bertrand (1887), P. Duhem (1911), L. Onsager (1931), C. Eckart (1940) and later, by J. Meixner, I. Prigogine, S.R. de Groot, P. Glansdorff and other authors is a great scientific success. In this connection, it was made possible, above all, to give a physical explanation of the formation of so-called dissipative structures (E. Schrödinger, A, Turing, L. v. Bertalanffy, I. Prigogine). It has to be stressed here again that such structures, however, do not simply result from the fundamental law (here, the thermodynamic laws), but mainly from a variety of assumptions on the specific character of the respective system (in particular, certain cooperative microprocesses and critical parameters are presupposed), that is, again from additional conditions. Accordingly, the origin of dissipative structures is described by bifurcations and corresponding perturbations.[165] It should also emphasized in this context that this physics cannot at all provide the theory of biological evolution; it cannot explain the biological evolution from primitive to complicated organisms. It is mutation, above all, that cannot be founded physically. From the view of physics, mutations can only be described as quantum-mechanical or thermodynamical fluctuations. As to the law, they are accidentally occurring changes in the initial conditions that is, a new setting of conditions in reproducing the biosystems.

Due to the fundamental meaning of time as quantity of measurement, all what was said about irreversible thermodynamics must not be understood as a revision of the physical space-time concept by which a kind of biological or even psychological time is introduced into physics. In particular, irreversible thermodynamics gives no reason for the assumption that the reduction of the time direction to the supplementary conditions, which specify the experimental situation and determine the respective solutions of the basic dynamic equations, would mean the introduction of subjective conceptions into physics. On the contrary, this reduction objectivizes the results of measurement by establishing a *consensus* among all observers under the same conditions. The observer loses his subjectivity, he can even be replaced by an automatic measuring device.[166] This view became especially clear in that version of quantum mechanics of systems with many particles (or degrees of freedom) that was elaborated by Rosenfeld together with George and Prigogine (the authors hold this extended version is the complementary and corresponding micro-theory or the non-linear irreversible thermodynamics): In that subdynamics of the extended quantum theory in which the

---

[165] See, e.g., I. Prigogine, *From Being to Becoming. Time Sciences*, San Francisco 1980.

[166] Cf. C. George, I. Prigogine, and L. Rosenfeld, "Macroscopic Level of Quantum Mechanics," op cit., p. 550.

macroscopic situation and thus the measuring device is described, the interaction between observer and measuring device does not make any contribution.[167]

In natural science, especially in physics, the subject remains explicitly only as an external observer and implicitly, it reflects itself in the *consensus* to consider the objective-real external world as given. In physical theory, the subject of cognition is represented by the means of cognition; it appears as measurement means in an objectivized manner.[168]

[167] Cf. ibid., p. 580.

[168] More details in this context are to be found in: E. Schrödinger, *Was ist ein Naturgesetz? Beiträge zum naturwissenschaftlichen Weltbild,* München–Vienna, 1962, pp. 33, 56, 62, 63, 72; *Mind and Matter,* Cambridge 1967, pp. 126-137; *Nature and the Greeks,* op. cit., pp. 90-98; see also H.-H. v. Borzeszkowski and R. Wahsner, "Erwin Schrödingers Subjekt- und Realitätsbegriff," *Dt. Zs. f. Philosophie,* vol. 35 (1987), 1109-1118; "Schrödingers subjektfreie Physik," in *Die Wirklichkeit der Physik,* op. cit., pp. 82-95; "Non-locality versus Locality: The Epistemological Background of the Einstein-Bohr Debate," op. cit.

Conclusion

A science of nature could only be founded by starting from the modern-age *(neuzeitlich)* standpoint on *motion as substance* because only from it the subject of physics was grasped as substance or, to use other words, only from it something was grasped as substance, which could be quantified without resulting in pure mathematics. To determine the *substance as (movable) body*, only admits counting or relations between numbers.[169] When the substance, the true, is seen in the motion, the cognition will not consist in the fact any more, as from the antique standpoint, to attribute a specific property to a specific object correctly but to show that an object *behaves* in accordance with its genus, that is, that an object moves. With regard to this, it is not possible to assume, as sometimes pretended, that antique philosophy of nature anticipated findings of modern natural science. Antique philosophy did pioneer work, of course; one example is the atomism mentioned above. It also thought the movable object – the movable individual "atom" – as substance and thus it was not able to link sensuousness and understanding, the sensual perception and the thought principle founded by it. Due to this atomism was not able to provide a sufficient epistemological basis for an empirical mathematized science of nature. But physics as a measuring science needs measuring bodies and respective theoretical presuppositions in order to conceive them, thus it needs bodies with only *one* behavior, it needs a principle to think isolated bodies. Distinguishing conceptionally between space and matter for the first time atomism formulated this necessary principle and developed the resulting consequences for the thinking of the contradiction constituting motion. For founding how the motion, thought by the appropriate distribution of momenta only existing in a unity, can be measured and what has to be considered as space and what as matter so that the motion can be measured, a conception had to be elaborated conceiving the motion as substance. Only this conception could produce the idea of the active principles of physics. Passive principles are properties of a substance grasped as a moveable object (in elaborated physical theories, they are firmly established to be presuppositions). But as Newton recognized explicitly,[170] physics cannot be constructed either on he basis of passive principles only - or on the basis of active principles only.

Physics takes the dialectical contradiction as dualism; this results from the fact that it has to determine and presuppose etalons in order to become a measuring science. Naturally, the etalons must not act on physical bodies (or: they may only act in a well-defined manner, under *one* well-defined aspect) and they must not be effected by them (or: they are only influenced by this *one* action). On the other hand, physical bodies are considered as mutually effecting. Of course, such a situation does not exist in nature and thus it has to be constructed in a experimentally and theoretically "arti-

169 See E. Cassirer, *Substanzbegriff und Funktionsbegriff*, op. cit.

170 See H.-H. v. Borzeszkowski and R. Wahsner, "Introduction" to *Voltaire, Elemente der Philosophie Newtons/ Verteidigung des Newtonianismus/ Die Metaphysik des Neuton*, op. cit.

ficial" way.[171] Therefore, one is not dealing with concrete bodies but with abstract ones to which a specific property or behavior is ascribed.

Thus, a measuring science will never be able to conceive the world in its totality; it can only be valid to the extent to which the abstraction made reflects an essential connection in a good approximation. It does not describe real but possible modes of behavior of real objects. This specific character of physics results in the fact that physical theory cannot grasp historicity without any mediation.[172] It was constructed just in such a manner that closed systems can be established that make it possible to define the measuring body "space-time" or the etalon of motion. The definition of all physical quantities is made via "conservation laws". A historical component, like it is necessary in cosmogony,[173] is introduced into physics by initial and boundary or other supplementary conditions which must always be determined when specific physical situations are to be calculated by means of general laws. The physical laws themselves do not contain this component; but they are compatible with it. This is the difference[174] between physical dualism and dialectical contradiction. The sides of dualism can also be determined only with regard to each other, they are also to be understood only in their relationship to each other, but they do not produce each other by their permanent conflict; but one side, the so-called passive principles, is presupposed and with respect to the considered physical theory determined *quasi-apriori* namely from the measurement-theoretical point of view. This is done by firmly establishing the result of the preceding conflict on the specific stage or under a specific aspect. By misunderstanding this basic situation of physics, by attempting, for example, to conceive the world in its totality on the basis of classical mechanics, the idea of historicity degenerates into the idea of a universal linear causality or into the Laplace demon. In principle, the same happens if one conceives the world in its totality on the basis of modern physical theories.

---

171 Thus, it is an error to think like Hegel that only philosophy – unlike other sciences – has not the advantage to admit their subjects immediately from the idea and to presuppose the method of recognizing for the beginning and continuation as already given. ["Die Philosophie entbehrt des Vorteils, der den anderen Wissenschaften zugute kommt, ihre *Gegenstände* als unmittelbar von der Vorstellung zugegeben, sowie die *Methode* des Erkennens für Anfang und Fortgang als bereits angenommen *voraussetzen* zu können. " (G.W.F. Hegel, *Enzyklopädie der philosophische Wissenschaften im Grundrisse (1830). Erster Teil. Die Wissenschaft der Logik,* op. cit., p. 41 (§ 1).)] For a discussion of this view of Hegel see literature cited in note 25.

172 See also respective considerations made by the French atomic physicist Pierre Jaeglé, in *Raum und Zeit. Grundfragen naturwissenschaftlicher und philosophischer Erkenntnis,* op. cit., especially pp. 48, 59-60, 88, 90, 93-97, 110 f. However, we do not hold, in contrast to Jaeglé, that taking this aspect of physics into account, it would not be possible any more to determine the dialectics as a theory of historicity. We hold it necessary to study more profoundly the relationship between physics and dialectics, the relationship between a calculating and measuring science, on the one hand, and philosophy, on the other.

173 So-called evolution models of the cosmos understand evolution as the expansion of the universe that does not differ from the spatial motion in its status. as far as such models describe the history of universe, this results essentially from supplementary assumptions and not so much from physical laws. (Cf. H.-H. v. Borzeszkowski and R. Wahsner, "Evolutionism as a Modern Form of Mechanicism," *Science in Context,* vol. 2 (1988), 287-306.)

174 Cf., note 1.

The entire forming of concepts and theories in physics is essentially determined by measurement requirements.[175] Thus, it is not possible to discuss about problems of the forming of physical concepts or theories or about physical truth by abstracting from measurement and its requirements. Without measurement, the subject of cognition cannot be mediated with the object of cognition. Thus, its possibility must be presupposed in each theory. This possibility of presupposition is not identical with theoretical performances in advance *per se* on which the respective theory is based. Certainly, all theories are based on preceding theoretical findings. But, in general, they do not mediate the subject and object of cognition. To see measurement only as the reading of positions of pointers, as the comparison with given standards or as calculable observations of natural connections of action in a calculable manner, cannot result in understanding its outstanding position in the process of physical cognitions. The fact has to be taken into account that measurement always presupposes quantities to which belong also rules saying how to measure these quantities, how to compare the measurements made at different space-time points and how to think the realization of etalons theoretically physically. Thus, a physical theory must always be formed in such a manner that these demands are satisfied. Physics mathematically formulated differs from mathematics by quantities of measurement and the consequences arising from their application. To objectify (*vergegenständlichen*) the equalities grasped by the respective quantities, the real measurement requires the artificial production of ideal situations; it requires objects that were prepared by real idealization, that is, by the purposeful exclusion of certain influences and effects, in such a manner that they can be used as objectified measures (e.g., as clock or measurement rod). In experiments, one operates with ideal and not with concrete bodies under ideal and not under concrete conditions. Both bodies and the conditions must be produced but one also must be able to think (not only to imagine) them, to think them in the respective theory. The experiment is a method to realize (purely) equalities and their relations. Only through this mutual approaching of theory and actuality, measurement can become mediation. The fact that measurement-theoretical determinations must be both presupposed and attached to dynamics was called by us "dualism".

The fact that physics excludes historicity in the manner explained above does not mean at all that it is placed besides the philosophy (besides a dialectical philosophy that overcomes the deficiencies of the Hegelian one) without mediation but on the contrary, it indicates the necessity of philosophy for physics or for the empirical mathematized natural science as a whole. This is true, because there is no natural science that is able to explain *per se* its epistemological basis. For this purpose, it does not suffice to consider any philosophy but it requires a philosophy founded on a conception of objective sensuousness (*gegenständliche Sinnlichkeit*). This is the only basis on which it is possible to investigate and elucidate the origin of thought determinations presupposed to physics ad the genesis of the thought entities of natural science as well as to explain how the linkage between these thought entities and real natural objects is established.

175 This does not mean, of course, that this establishing is measurement or that measurement *per se* would produce concepts or theories.

The calculable and measurable contradiction can never be, in its essence, the philosophical-concrete, the dialectical contradiction. However, the dialectical contradiction namely the contradiction not contained in the notion for itself cannot be fully comprehended if the thought has not taken the way through this calculable and measurable contradiction and thus through the physical dualism.

Appendix
on Physical Mechanics and Philosophical Mechanicism

Voltaire's Newtonianism
A Bridge from English Empiricism to Cartesian Rationalism and Its Implications for the Concept of Mechanics in German Idealism

Horst-Heino v. Borzeszkowski and Renate Wahsner

"My fate has decreed that I should be the first of my countrymen to be permitted to explain the great Newton's discoveries. I have been the apostle and the martyr of the English"[1] – claimed Voltaire, celebrated beacon of the French Enlightenment. His aim, according to his own account, was "to comprehend Newton and to render him comprehensible".[2]

It was Voltaire who unleashed the spread of newtonianism in France. Certainly, the Paris Academy was aware of all Newton's writings from 1699, and in the period which followed there were other Frenchmen – Maupertuis, d'Alembert, Clairaut, Lagrange and Laplace among them – who developed classical mechanics further, but it was Voltaire who really popularized these works in French. In so doing, he prepared the ground for the course which mechanics were later to pursue. However, his efforts did not only exert an influence on the scientific, but also on the philosophical and, indeed, cultural developments in general. By propagating Newton's findings as a world view which ultimately challenged the altar and the crown in an absolutist state, he struck a decisive blow for justifying the claim of reason, helping to establish it as the sole criterion for all activity, thereby paving the way intellectually for the French revolution and the progress of Enlightenment in France.[3]

Newtonianism would not have been accepted in France in its English form. Born essentially from an empirical attitude, English newtonianism had to be dressed for the

---

1 "Ma destinée a encor voulu que je fusse le premier qui ai expliqué à mes concitoiens les découvertes du grand Neuton, que quelques sots parmi nous appelent encor des sistêmes. J'ai été vôtre apôtre et vôtre matir. En vérité il n'est pas juste que les Anglais se pleigneut de moi." (Voltaire, "Letter to Horace Walpole of 15 July 1768," in *Correspondence and Related Documents,* definitive edition by Theodore Besterman, in T*he Complete Works of Voltaire,* ed. by W. H. Barber and Ulla Kölving, vols. 85-135, Institut et Musée Voltaire, Genève 1968-1977, vol. 117, p. 449 (D 15140). – All letters therein are published in the original language.)

2 "Du reste je vis assez en philosofe, j'étudie baucop, je vois peu de mondes, je tâche d'entendre Neuton et de le faire entendre, je me console avec l'étude, de l'absence de mes amis; il n'y a pas moyen de recoudre àprésent l'enfant prodigue." (Voltaire, "Letter to the Count d'Argental of 27 January 1737," in *Correspondence and Related Documents,* op. cit., vol. 88, p. 222 (D 1270).)

3 The present essay is essentially a synthesis of the following works: H.-H. v. Borzeszkowski and R. Wahsner, "Die Mechanisierung der Mechanik. Rezeption und Popularisierung der klassischen Mechanik durch Voltaire," in *Newton-Studien,* Berlin 1978; *Newton und Voltaire. Zur Begründung und Interpretation der klassischen Mechanik,* Berlin 1980; "Introduction" to *Voltaire, Elemente der Philosophie Newtons/ Verteidigung des Newtonianismus/ Die Metaphysik des Neuton,* edited, introduced and supplied with an Appendix by Renate Wahsner and Horst-Heino v. Borzeszkowski, Berlin 1997; "Not Even Classical Mechanics is Mechanistic," in *Worldviews and the Problems of Synthesis,* vol. 4: The Yellow Book of "Einstein Meets Magritte", ed. by D. Aerts, H. van Belle, and J. van der Veken, Dordrecht–Boston–London 1999, pp. 251-262.

Cartesian rationalism which prevailed in France. Voltaire found the recipe, and in the process he opened up traditional philosophy, understood as rational metaphysics, to problems which hitherto had lain within the territory of empiricism. Inherent in Voltaire's success, however, there was a mechanistic interpretation of Newton's teaching, which after all had not been empiricism as such, but an empirical natural science expressed in a mathematical language.[4] When Voltaire adopted Newtonian physics as a world view, he was ignoring its conditionality and limited nature. Voltaire initiated a philosophical reading of mechanics which subsequent generations of philosophers regarded as the only reading possible, rather, mistaking it for mechanics pure and simple. Specifically, the concept of mechanics which came to underlie German idealism reflects Voltaire's reading of Newton. There is a twofold implication to this: classical mechanics is identified with the mechanistic world view, and Newton's principles are scrutinized for any *cultural* insights they might have to offer. The impact of Voltaire's perspective remains with us today, even if the second side to this coin has been much neglected.

Voltaire's Little Catechism on Gravitation

Voltaire's writings on English philosophy and Newtonian physics [specifically his *Philosophical Letters* (1733), the *Eléments de la philosophie de Newton* (1738/41) and *Défense du newtonianisme* (1739)] attracted considerable attention. His accom-

[4] To recognize this one must, first, be able to conceptualize the difference between natural science and philosophy [cf. H.-H. v. Borzeszkowski and R. Wahsner, "Über die Notwendigkeit der Philosophie für die Naturwissenschaft," in *Dialektik 1. Beiträge zu Philosophie und Wissenschaften. Orientierungen der Philosophie*, ed. by B. Heidtmann, Cologne 1980; "Noch einmal über das Bedürfnis der Naturwissenschaften nach Philosophie," in *Dialektik 5. Beiträge zu Philosophie und Wissenschaften. Darwin und die Evolutionstheorie*, ed. by K. Bayertz, B. Heidtmann and H.-J. Rheinberger, Cologne, 1982] and, second, grasp that an action is not the same as consciousness of this action. In his *Principia*, therefore, Newton founds not only a physical theory in the modern sense, but also discusses problems relating to its epistemological status. The principles substantiated by Newton were later developed in more depth and detail, and Newton's geometrical formulation was transposed into analytical form [cf., for example, H.-G. Schöpf, "Newton zwischen Geometrie und Erfahrung," in *Gottfried Wilhelm Leibniz im philosophischen Diskurs über Geometrie und Erfahrung*, ed. by H. Hecht, Berlin 1991, pp. 162-185; R. Wahsner and H.-H. v. Borzeszkowski, *Die Wirklichkeit der Physik. Studien zu Idealität und Realität in einer messenden Wissenschaft*, Frankfurt a.M.–Berlin–Berne–NewYork–Paris–Vienna 1992, pp. 194-210]. In the process, Newton's action made its mark on philosophical consciousness, which enabled the difference between mechanics and the mechanistic world view to be recognized. Although this did not happen until the late 18th century, this does not alter the fact that the conception used in the *Principia* cannot be equated either with the ancient concept of τέχνη, or practical mechanics, or with the mechanistic world view [cf. H.-H. v. Borzeszkowski und R. Wahsner, "Die Natur technisch denken? Zur Synthese von τέχνη und φύσις in der Newtonschen Mechanik oder das Verhältnis von praktischer und theoretischer Mechanik in Newtons Physik," in *On the Culture of Morality. Practical Philosophy in the Post-traditional Society*, ed. by M. Weingarten, Berlin (in print) [also: Preprint 87 of the Max Planck Institute for the History of Science, Berlin 1998].

plishment was acknowledged even by some not especially well disposed towards him.[5] At the same time, a whole range of pamphlets and open letters expressed vehement criticism, not only from the Church, but also from most professional scientists.[6] The impact of Voltaire's *Elements* [7] can only be compared, some would maintain, with that of Darwin's *Origin of Species*.[8] To assess that claim, we must consider the situation in that period. Condorcet, an informed witness, reports: "When M. de Voltaire published this book, Europe's greatest mathematician Johann Bernoulli was still opposed to Newton's teaching; over half the *Académie des sciences* supported Cartesian thinking; even Fontenelle, so far removed from any sectarian or national prejudice and not yet thirty when Newton's system was published, and one of the few people capable of comprehending it, clung obstinately to his original views. If, in addition to all this, one considers that the first textbook for French schools to address Newton's theory did not appear until ten years after M. de Voltaire's work, one can only conclude that the publication of 1738, which our famous master so modestly called his little catechism on gravitation, deserves great credit."[9]

Philosophy's New Role

Voltaire's enthusiasm for Newton would have been inconceivable without his personal experience of English society. When he arrived in England in 1726, he was a philosophical novice, and had no firm views of a philosophical or religious nature. Confronted with religious diversity in England, he was compelled to think for himself

---

5 A Jesuit remarked: The great Newton seemed forgotten. At last M. de Voltaire appears, and at once Newton is understood, or at least people make an effort to understand him; all Paris resounds with Newton, all Paris stutters Newton, all Paris is studying and learning its Newton. (Cf. *Mémoires pour l'historie des sciences & des artes* (1738), 1673-1674; cf. also *Bibliothèque française XXVIII* (1738?), 257-289.)

6 Some of the latter because they still adhered to pre-Newtonian opinions, and others because they found Voltaire's account betrayed the amateur.

7 First published in English in 1738 as: *The Elements of Sir Isaac Newton's Philosophy*. By Mr. Voltaire. Translated from the French. Revised and Corrected by J. Hanna, M. A. London. As this edition does not contain the "metaphysical" chapter, in the following, ciphers in square brackets are related to the separate edition *The Metaphysics of Sir Isaac Newton: or, A Comparison between the Opinions of Sir Isaac Newton and Mr. Leibnitz.* By M. de Voltaire. Translated from the French. By David Erskine Baker, London 1747.

8 Cf. Th. Besterman, Voltaire, London and Harlow 1969, p. 193.

9 "J'aurais pu faire observer encore que lorsque m. de Voltaire donna cet ouvrage, le premier des géomètres de l'Europe, Jean Bernoulli, combattait encore le neutonianisme; que plus de la moité de l'Academie des sciences était cartésienne; que Fontanelle enfin, si supérieur à tous les préjugés de secte ou de nation, Fontenelle, qui n'avait pas trente ans lorsque le système de Neuton parut, & qui était du petit nombre de gens qui pouvaient l'entendre; que Fontanelle était resté opiniâtrement attaché à ses premièrs opinions. Si on ajoute à tout cela que le premier livre classique où l'on ait développé en France les théories de Neuton, ne parut que dix ans après l'ouvrage de m. de Voltaire, on ne peut se dispenser de convenir qu'il y avait bien du mérite, en 1738, à donner ce que notre illustre maître appelle avec tout de modestie son petit catéchisme d'attraction." (M.J.A.N. de Caritat Condorcet, "Letter to J. F. de La Harpe of June 1774," in *Voltaire, Correspondence and Related Documents*, op. cit., vol. 125, p. 24 (D 18991).)

and discovered the protective benefits of tolerance. Moreover, tolerance in practice opened up new vistas to him, beginning with those of religious debate. He talked at length with Quakers, Jews, and also the vicar Samuel Clarke, who dreamed of proving god's existence by means of mathematics. He made the fleeting acquaintance of Bishop Berkeley, and was much inspired by Bacon. Blessed with the art of conversation and a lively intelligence, he was able to compete with learned theologians. A grandiose prospect unfolded before him: unbridled communication between the world of scientists and the world of writers. Voltaire was thus able to reveal his one of its most brilliant talents: the ability to grasp a fact going out of ordinary and transform it into a myth (as with the apple falling on Newton's head). In effect, Voltaire helped force the ruling caste, or public opinion, to take note of science.

Alongside tolerance, Voltaire discovered prosperity and industriousness in England, and he combined these various discoveries, broadening the sphere of literary intervention to include anything of interest to society. His adage was that there were no forbidden fruits for the writer. A field should be closed to literature neither because it concerned a science allegedly too complex or scholarly, nor because it concerned secret affairs of state. The writer's new role should be to voice the needs and rights of civil society and, at the same time, to bring it education.

Nevertheless, Voltaire wished to preserve literature within the sphere of noble activity, or perhaps to raise it to that status. Encouraged by the tribute which England paid – as Voltaire himself witnessed – to the scholar and writer Newton at his funeral, a tribute which France only granted to kings, Voltaire was convinced that this objective was attainable. He returned from England with a taste for liberty, tolerance, modernization, utilitarianism and the involvement of writers in public affairs, and also a mission to apply reason and a critical spirit. He wished to fight for an organization of society founded upon the universal implementation of trade, and yet he resisted the incorporation of literary activity and artistic relations into that trade structure.

As the citoyen began to enter the social stage, the writer acquired a function whereby conversation was no longer a pleasant art, but the most perfect manifestation of this new social being, this new force of opposition on the social stage who now confronted the exclusive power of the state. But the "free" writer, emancipated from church and court, was now dependent on a new master: the public.

Voltaire approved of the principles underlying this new reality, but he feared their consequences, and for this reason he often resorted to contradictory tactics. He did not wish to choose between two camps which seemed to him equal in desirability: loyalty to the past or the fascination of a new society, respect for the elite or celebration of the mass, nobility or money, refined taste or vibrant sensibility. Voltaire plunged wholeheartedly into the buzz of challenge so typical of this new public arena. The motto was critique on every front: the critique of science, of history, of religion, of politics. And another name for this all-round critique was "philosophy".

Since returning from England, Voltaire had made notes on his *English Letters* (later known as the *Philosophical Letters*). They were originally intended merely as observations on England's delightful modernity. In Voltaire's own judgment, however, their great strength lay in their philosophical component.[10] Voltaire recognized a need on the part of the society in which he lived, or, rather, something his society required if it was to be transformed, but barely possessed. Increasingly conscious that his stay in England had changed him, he interpreted the hysteria which gripped high society in France as a yearning for change – change which he also feared. As a remedy, he proposed a literary genre old in content but new in form: philosophy.

This contrast between content and form reflected the contention that philosophy in the guise of learned science would not solve any problems of society, that Gassendi, Descartes and Newton had each bequeathed a clear message, but that society had no knowledge of it because it was formulated too earnestly and decoding it demanded too much effort. *To enter people's minds, science must become literature* (called philosophy), conceding a role for charm, frivolity and worldliness. (Fontenelle had shown the way, but he had been too cautious and too harmless to threaten power with anything more than a mild disturbance. In 1727, for example, he as an acknowledged and respected personage had delivered a speech on Newton's death in homage to the author of the *Principia*, seeking to defend the Cartesian world view without discarding Newton's discoveries.[11]) For Voltaire, the consequence was to place talent in the service of truth, convinced that he was satisfying the new needs of his public. The tremendous success of his *Philosophical Letters* seemed to prove him right, and it also brought him recognition as a philosopher,[12] a friend of wisdom and thereby of truth.[13]

Philosophy, the exploratory quest for truth, was primarily a movement to combat the influence of prejudices which barred access to knowledge and obstructed its efficacy. Philosophy as Voltaire and his contemporaries understood it was only conceivable as a rival to religious propaganda. It could only hope to win people over by occupying terrain where the church had been entrenched for centuries. Philosophy had to fight above all on the territory occupied by religion: knowledge, education, emotions, morality, politics. To fulfill their task, these philosophers had to claim that, like priests, they were "anointed" to promote the happiness of a human race which could achieve harmony with itself and with the truth about the objective world on the basis of reason. The philosopher was the new priest of a human religion, a religion for humanity released from original sin, seeking the secret of greatness and dignity within himself.

---

10 Cf. Voltaire, "Letter to Jean-Baptiste-Nicolas Formont of June 1733," in *Correspondence and Related Documents*, op. cit., vol. 86, pp. 342 f. (D 617).

11 Cf. B. Fontenelle, "Laudatio auf Newton," in B. Fontenelle, *Philosophische Neuigkeiten für Leute von Welt und für Gelehrte. Ausgewählte Schriften*, Leipzig 1989, pp. 326-348.

12 20.000 copies were apparently sold between 1734 and 1739.

13 For a detailed background to the publication and impact of the *Letters* see, for example: T. Besterman, Voltaire, op. cit., pp. 110-124, 171 f.

Voltaire's Fascination with Newton

Voltaire was enthusiastic about scientific perceptions of his day. "A new universe was discovered by the philosophers of the last century", he mused, "one that was all the more difficult to understand as people did not even suspect that it existed. The wisest felt that there was some temerity in so much as dreaming that one could divine the laws by which the heavenly bodies move and by which light manifests itself."[14] Newton had not merely dreamt of divining, but had formulated answers to many questions which absorbed thinking minds at the time: Why do planets and comets follow a particular path and what holds them there? Why do bodies fall to the earth? What causes the tides? Newton answered these questions, and what is more – according to Voltaire – by applying a single principle: "*attraction,* the great means by which all nature is moved",[15] he explains in his *Philosophical Letters*, while in his *Dictionnaire Philosophique* he says of Newton: "He has discovered and demonstrated a new principle which causes all movement in nature."[16]

In other words, Voltaire saw Newton's law of gravitation (a specific law of a specific theory pertaining to a particular science) as a principle which explained the whole world. The kind of world law which he believed to have been demonstrated implies, first, that everything which happens does so of absolute necessity and, second, that there must be a lawgiver and a motor. According to Voltaire, everything is subject to one law. Countering the objection that nothing therefore exists but machines, he argued: "Well, would you have everything rendered subject to a million of blind caprices? Either all is the consequence of the nature of things, or all is the effect of the eternal order of an absolute master; in both cases we are only wheels to the machine of the world."[17] Voltaire relates this omni-explanatory principle not only to nature, but also to human beings. Taking this natural law to be universal, he uses it to substantiate the absence of free will, and thus the impossibility of arbitrary human action. His axiom: We may do what we want, but we cannot want what we want.[18] In a letter to the Marquise du Deffand he writes: "Every being, every kind of existence is dependent of necessity on the world law. It is ridiculous, we are told, and impossible for man to be able to give himself something if the host of heavenly bodies can give itself nothing. How can we be free ... if the universe is a slave? ... All our losings, all our emotions, all our thoughts are absolutely necessary things."[19] When Voltaire states that

---

[14] Voltaire, *Philosophical Letters*, translated, with an Introduction, by Ernest Dilworth, Indianapolis–NewYork–Kansas City 1961, p. 75.

[15] Ibid., p. 72.

[16] "... il a découvert et démontré un principe nouveau qui fait mouvoir toute la natur." (Voltaire, "Newton et Descartes," *Dictionnaire Philosophique*, in *Œuvres complètes*, Paris 1831, vol. 34, p. 332.)

[17] Voltaire, "Free-will," in *Voltaire's Philosophical Dictionary*, ed. by W. F. Fleming (Ten Volumes), Paris–London–NewYork–Chicago 1901, vol. V, p. 131.

[18] Cf. ibid., pp. 128-132; Voltaire, *Élémens de philosophie de Newton*, in *Œuvres complètes de Voltaire*, Tome trente-unième, De l'imprimerie de la société littéraire-typographique 1784, pp. 38-51. [19-28.]

[19] "Chaque être, chaque manière d'être obéit nécessairement à la loi génerale. Il est ridicule et impossible que l'homme se donnât quelque chose, quand la foule des astres ne se donne rien, et que

everything – from a blade of grass to an oak tree, from a flea to a man, from the grain of sand up to our clouds – is a spring, lever and roller", a hydraulic machine or a chemical laboratory,[20] he is not simply motivated by the pleasure he always derived from mocking common assumptions, but expressing his view that the law of the world is a universal mechanism to which everything which happens is slavishly subsumed. He sees the world as a machine which functions according to a pre-ordained law with no exceptions. There is no such thing as accident. In this manner, a world law amounts to exactly the same as divine providence.[21] To this end, however, Voltaire needs a god as lawgiver and as the workman who builds the world according to plan, a god as learned as the Royal Society in London.[22] But for Voltaire's world machine to work, it also needs a motor. Voltaire's god must perform this function, too. Without this motor, matter cannot be either ordered or brought to life. For, "how can matter have motion by itself, as it has, according to all the ancients, extent and divisibility?"[23] In answering this question, Voltaire does not commit himself totally, although he thinks it more likely that an incorporeal cause created the motion. His approach would not have permitted him to substantiate the opposite view.

Voltaire believed that all this was implicit in Newton's mechanics. He fell prey to this misunderstanding because he was unaware of the reason why Newton spent his life wrestling with the phenomenon of gravitation and the character of physical laws. Although he knew Newton's *Principia* and the Questions on the *Opticks* (in *Eléments de la philosophie de Newton* he specifically refers to both), he evidently did precisely what Newton feared readers would do if they "lacked sufficient insights into the principles".

Newton's Mechanics as a Measuring and Calculating Science

What worried Newton about readers of this kind? In the Introduction to Book Three of his *Principia*, his book on the "System of the World", he writes: "In the preceding

---

nous fussions libres (dans le sens des théologiens) quand l'univers est esclave. ... Toutes vos privations, tous vos sentiments, toutes vos idées sont des choses absolument nécessaires." (Voltaire, "Letter to the Marquise du Deffand of 22 May 1764," in *Correspondence and related Documents*, op. cit., vol. 111, p. 387 (D 11883).)

20 Cf. Voltaire, "The Clergyman and His Soul," in *The Writings of Voltaire, Four Volumes in One*, NewYork City 1931, p. 18. ["Animaus, végétaux, minéraux, tout me paraît arrangéavec poids, mesure, nombre, mouvement. Tout est ressort, levier, poulie, machine hydraulique, laboratoire de chimie, depuis l'herbejusqu'au chêne, depuis la puce jusqu'à l'homme, depuis un grain de sable jusqu'à nos nuées." (Voltaire, "Les Oreilles du Comte de Chesterfield," in *Œuvres complètes*, Paris 1830, vol. 37, p.126 ; see also: *Élémens de philosophie de Newton*, op. cit., p. 55. [35 f.])]

21 Cf., for example, Voltaire, "Zadig," in *Œuvres complètes*, Paris 1830., vol. 36, pp. 75 f.

22 Cf. Voltaire, The Clergyman and His Soul, op. cit., pp. 17 f. ["Il faut assurément que quelqu'un d'aussi savant que la Société royale Londres ait arrangé les choses de manière que le carré des révolutions de chaque planète soit toujours proportionnel à la racine du cube de leur distancea leur centre; et il faut être sorcier pour le deviner." (Voltaire, "Les Oreilles du Comte de Chesterfield," op. cit., p. 126.)]; cf. also "Atheist," in *Voltaire's Philosophical Dictionary*, op. cit., vol. II, p. 130.

23 "Matter," in *Voltaire's Philosophical Dictionary*, op. cit., vol. VI, p. 245.

Books I have laid down the principles of philosophy; principles not philosophical but mathematical: such, namely, as we may build our reasonings upon in philosophical inquiries. These principles are the laws and conditions of certain motions, and powers or forces, which chiefly have respect to philosophy."[24] It is significant to talk of building, for this implicitly casts light on how Newton understood physics. From the very outset, the definitions with which he begins his *Principia* lend his theory (expressed in mathematical terms) a physical significance (that is, one which relates to reality) by defining the basic concepts of mechanics (presented as physical quantities) as dispositions for measurement. These definitions are followed by the laws of motion – Newton's three laws – which establish mathematical links between these physical quantities. These links derive from propositions which have been obtained in the course of a complicated trajectory by which previous experience (above all that reflected in Kepler's and Galileo's laws) has been subjected to theoretical treatment. Now that Newton has formulated the laws of motion, he demonstrates mathematical propositions in Book One which he then uses in Books One and Two to derive mathematical conclusions about centrosymmetric forces and other situations (including motion subjected to various retarding forces), and he expresses these in the form of theorems. By and large these are calculations which he carries out as generally as possible, without reference as yet to any specific physical instances. It is only in Book Three that he then applies many of the theorems he has derived to the specific instance of gravitational motion by the moon, the planets, the comets and the sea.

Anybody unaware of the role played by mathematics and experimentation, and unable to follow the ideas in the first two books due to a lack of knowledge and practical experience in the fields of mathematics and physics, would not really be in a position to grasp the *Principia* as a whole. This reader would only understand a few of the more general passages in the first two books and a number of pages in Book Three. Evidently, the scope for misinterpretation was, therefore, considerable. Newton, at least, had suspected that this might be a danger, and it had prompted him to drop his original plan of publishing a popular version of Book Three, hoping that anyone incapable of comprehending the earlier parts would be deprived of the opportunity to read the third. "I had, indeed," he relates, "composed the third Book in a popular method, that it might be read by many; but afterwards, considering that such as had not sufficiently entered into the principles could not easily discern the strength of the consequences, nor lay aside the prejudices to which they had been many years accustomed, therefore, to prevent the disputes which might be raised upon such accounts, I chose to reduce the substance of this Book into the form of Propositions (in the mathematical way), which should be read by those only who had first made themselves masters of the principles established in the preceding Books."[25]

Newton, then, explicitly maintained that anyone seeking to prise a handful of verbal sentences from the overall fabric of the work would fail to understand his physics, or

[24] *Sir Isaac Newton's Mathematical Principles of Natural Philosophy and His System of the World,* translated into English by A. Motte in 1729, the translations revised, and supplied with an historical and explanatory appendix, by F. Cajori, Berkeley (California) 1947, p. 397.
[25] Ibid.

his "Mathematical Principles of Natural Philosophy", for, his mechanics expressed natural laws in the form of mathematical equations connecting physical quantities and was not viable without this mathematical language or without empirical foundations. Naturally, his unambivalent declaration did not prevent some people who had not mastered the preceding principles from reading the book, and thus the matter was sucked after all into dispute, a dispute which has essentially persisted until today. One of the core issues is the debate about the mechanical concept of force and Newton's law of gravitation.

A New Concept of Force

In Newtonian mechanics the concept of force is essentially determined by means of the three laws of motion, and in particular the second, which states that *mass* × *acceleration* = *force*, or $m\mathfrak{b} = \mathfrak{F}$. A special term of force is still required for the symbol $\mathfrak{F}$ in order to express a dynamic interaction. To obtain the law of gravitation, therefore, the term of force for gravitation must be inserted on the right-hand side of Newton's second law of motion. This then reads: $m\mathfrak{b} = -G\frac{M \cdot \mu}{r^2}\mathfrak{r}_0$ (where m is the inertial mass, $\mathfrak{b}$ the acceleration, G the gravitational constant, M and μ the heavy masses, r the distance between them and $\mathfrak{r}_0$ the unit vector pointing from the mass M to the mass μ).

Voltaire, by not appreciating the significance for mechanical statements of Newton's three laws or, indeed, of any mathematical-physical principles developed in the first two books, also failed to realize that no statements can be made about the mechanical concept of force in isolation from these principles. To him, the concept of force was mainly sensuous. Force was anything which induced movement.[26] When he does try to define force he does not hit upon the equation $m\mathfrak{b} = -G\frac{M \cdot \mu}{r^2}\mathfrak{r}_0$ which we find in Newton's theory,[27] but is misled by the formulations of Leibniz and Descartes to see the expression $\frac{m}{2}v^2$ or mv as the fundamental concept of force in mechanics.[28] Rather than discussing the law of gravitation in the form of the above equations which determine gravitational interaction, he describes how the force of gravitation is determined by masses and the distance between them. In other words, the equations which

26 Cf., e. g., Voltaire, *Élémens de philosophie de Newton*, op.cit., pp. 75-78 [65-72].

27 Cf course, in Newton's *Principia*, one does not find this equation in this analytic form but in a corresponding geometric formulation. Sometimes this fact leads to the curious misunderstanding that *time* does not occur in this theory. This view ignores that Newton defines a *change of motion* as change per time-interval. Possibly, the misunderstanding arises because Newton does not mentioned this explicitely, and for people who do not follow his deductions this might be difficult to see. (For the relation between the geometric and alalytic formulation in Newton, see, e.g., H.-G. Schöpf, "Newton zwischen Geometrie und Erfahrung," op. cit.)

28 Cf. ibid., pp. 75-78.

constitute the law of gravitation are reduced to their right-hand side. To put it another way, Voltaire equates the law of gravitation with the force term for gravitation.

This explains why Voltaire could not ultimately admit other forces than those of gravitation. Unlike Newton, he could not simply insert any other force into the right-hand side of the equations $m\mathfrak{b} = \mathfrak{F}$, but would have needed to introduce a new principle, thereby thwarting his own goal of explaining the world in terms of a single principle.

When Voltaire talks of "force" as the fundamental principle of mechanics, he concurs verbally with Newton, but in substance there is a great difference. It is true that, in Newton's theory, nature is explained by the effect and interplay of diverse forces, and since these are defined by the gravitational equations described above, Newton's second law of motion basically encompasses the full dynamics of classical mechanics, but it does not embrace force as something sensuous-illustrative or as something determined by life experience.

If the basic principle of mechanics is equated with the law of gravitation and this with the term for the gravitational force, or rather a diffuse notion of force, and all of this at once with the law of the whole world, then to some extent the reason – or rather the trigger – for this conflation is rooted in mechanics itself.

Although the dynamics of gravitation is merely a specific instance of Newton's universal dynamics, it also plays an outstanding role in history in that their development was a synthesis between experimental research into earthly mass (above all by Galileo) and theoretical astronomy (primarily formulated by Kepler). The law of gravitation was the answer to a question much discussed at the time about whether bodies fell to the ground for the same reason that planets remained within their orbit. Following the Copernican principle, Newton's gravitational dynamics assumed the unity of natural law, so that the same natural laws were at play on earth as anywhere else in the universe, and he proved it. Nature now embraced heaven and earth, things and phenomena were universally comparable – an idea pursued since the dawn of the Modern Age, but now realized for the first time in a scientific theory.[29] Gravitation made this possible because it takes effect universally, over distances of both terrestrial and cosmic magnitude. That is why it was easy to regard the law of gravitation as the law of the world. In this context, unlike Newton, one often understood "world" not as the solar system or as the system of fixed stars but in a philosophical sense.

Due to the exceptional role of gravitation, it is sometimes claimed that all Newton did was apply mechanics to gravitation. This overlooks the fact, however, that mechanics as Newton knew it was essentially earthly mechanics, and could not be applied to heavenly bodies, and thus to gravity, until Newton had synthesized it with Kepler's astronomy (and other contributions from physics, such as Huygens' hydrodynamics), as a result of which mechanics was no longer what it had been, but a theory of dynamic interaction.[30]

Although Newton himself had a great many questions about the relationship between physics and mathematics, having devised the mathematical language of mechan-

[29] Cf. R. Wahsner, *Mensch und Kosmos. Die copernicanische Wende*, Berlin 1978, "Introduction".
[30] Cf. F. de Gandt, *Force and Geometry in Newton's Principia*, Princeton 1995.

ics he knew, unlike Voltaire, that there was no adequate non-mathematical translation for it. That is why, even when Voltaire repeats Newton's statements almost word for word, he does not necessarily mean the same thing. Inevitably Voltaire did not understand why the very thing which repeatedly drove Newton to ponder upon the essence of gravitation was the thorny problem of the extent to which a mathematical equation can express a law of natural physics.

In his *Principia* Newton takes the view that, although he has described gravitation mathematically, he has not explained it physically: "I likewise call attractions and impulses, in the same sense, accelerative, and motive," he writes in his Definitions for the work "and use the words attraction, impulse, or propensity of any sort towards a centre, promiscuously, and indifferently, one for another; considering those forces not physically, but mathematically."[31] Although he was fully aware – as we have seen – that the insights derived from his work in physics could not be prised out of the mathematical structure in which they were expressed, he was nevertheless dissatisfied with this state of affairs and sought an additional – so he believed – physical explanation. His dissatisfaction was rooted in the fact that Newton, as the founder of theoretical physics, had created something new, but that naturally the full consequences could not be perceived at the moment of birth, all the less the consequences for the relationship between physics and mathematics and for the concept of physical explanation.

A "physical explanation" was understood in Newton's day to mean establishing the origin of phenomena in terms of the characteristics of individual atoms, such as extension, hardness, indivisibility, mobility and inertia. For this reason Newton was constantly at pains to discover the physical cause of gravitation by using ether models. At the same time, his law of gravitation told him that gravitation is not the property of an individual atom or body, but that bodies are only heavy towards each other. Gravitation, then, cannot be reckoned among the primary characteristics of the smallest particles. *That* is why Newton always objected vigorously to misinterpretations of his message: "You sometimes speak of Gravity as essential and inherent to Matter. Pray do not ascribe that Notion to me; for the Cause of Gravity is what I do not pretend to know, and therefore would take more time to consider of it."[32]

In the 31st of the Questions which accompany his *Opticks*, after debating various other solutions for some time, Newton had the idea that gravitation was an active principle (unlike inertia, which was passive). An active principle, to him, was not an "occult quality", but a general law of nature "by which the Things themselves are form'd",[33] as an indissoluble tension, a behaviour which existed only in the action of bodies "towards each other", and which could not be reduced to the property of a atom or body as a single. By introducing the concept of an active principle, "physical

[31] *Sir Isaac Newton's Mathematical Principles of Natural Philosophy*, op. cit., p. 5.

[32] Cf. I. Newton, "Letter to R. Bentley of 17 January 1692/3 (2nd letter)," in *Isaac Newton's Papers and Letters on Natural Philosophy*, ed. by I.B. Cohen, Cambridge, Massachusetts 1958, p. 298; *Sir Isaac Newton's Mathematical Principles of Natural Philosophy*, op. cit., pp. 400, 546 f. (Regulae philosophandi, Scholium Generale).

[33] Cf. I. Newton, *Opticks*, with a foreword by A. Einstein, an introduction by Sir Edward Whittaker, a preface by I.B. Cohen, Dover 1952, p. 401.

explanation" no longer means to reduce phenomena to primarily atomic properties but to laws of nature which embraced physical interaction. This determined both the necessary mathematical formulation and the acknowledgment that natural bodies have the ability to act, and the result was a new concept of physical reality.[34]

Voltaire was well aware of Newton's reservations. Indeed, he quotes them in his own writing, although in such a way that, in defiance of Newton's concept, he actually manages to make a case for gravitation as an inherent force. Hence Voltaire was able to write in his *Philosophical Letters* that Newton regarded attraction as the great prime mover by which all nature is moved, a central force found in all bodies.[35] He attributes to Newton the argumentation that: "I use the word *attraction* only to express an effect that I have discovered in nature: the certain and indisputable effect of an unknown principle, a quality inherent in matter, of which cleverer men than I will find the cause, if they can."[36]

If Newton tussled to comprehend a novel theory, Voltaire reflected his efforts with a mixture of agnosticism and a conviction that he had the world firmly in his grasp. Newton, he maintains, had foreseen resistance to "the mere name" of this great and fundamental principle of attraction: "In more than one place in his book he cautions his reader about attraction itself, warning him not to confuse it with the occult qualities of the ancients, and to content themselves with the knowledge that in all bodies there is a central force which, from one end of the universe to the other, acts on the nearest bodies and on the most remote, according to the immutable laws of mechanics."[37]

By attributing properties to individual atoms or bodies in this manner, properties which they can only possess in their relationship with one another or, rather, their behavior towards one another, Voltaire cannot avoid being a little mystical about motion, for he cannot explain where it comes from. His stance also renders any mathematical formulation of gravitation (or other active principles) superfluous, for the *relationship* expressed in the mathematical terms of the law is regarded as something inherent in the separate bodies. The mechanical concepts are thereby prised out of the mathematical structure, the only matrix in which they have a specific meaning, and treated as if they had that meaning regardless.

By interpreting gravitation as a central force which is *in* each separate body, Voltaire ignores the ability of bodies to interact. It is of no matter to him that bodies are only heavy *towards each other*. He thinks of the dynamic interaction known in a physical theory as "force" as an impact. Having assumed that force is inherent in separate pieces of matter, Voltaire imagines that to each gravitating body belongs a central point to which it must be striving towards. However, as geometric points cannot exert a physical force, Voltaire must have recourse to god as the cause of motion. "If matter

---

[34] Cf. also R. Wahsner, *Das Aktive und das Passive. Zur erkenntnistheoretischen Begründung der Physik durch den Atomismus – dargestellt an Newton und Kant*, Berlin 1981, Chapter I and II; R. Wahsner and H.-H. v. Borzeszkowski, *Die Wirklichkeit der Physik*, op. cit., pp. 29-96, esp. 29-47.

[35] Cf. Voltaire, *Philosophical Letters*, op. cit., p.72.

[36] Ibid., p. 73.

[37] Ibid., p. 72.

strives towards a points, as has been proven, it nevertheless does not appear to gravitate of its own accord, as it acquires its extent from nature. It has acquired its gravity, therefore, from god."[38] "He has decreed a law for all bodies by which they must all strive in the same way towards their central point."[39] Thus Voltaire separates matter from motion; matter acquires its motion from outside.

Newton and Voltaire on the Limitations of Mechanics

As long as Voltaire applied his blueprint, god or some other principle, external to physics, essentially had to keep pumping motion into the world, accomplishing a function that Newtonian physics demonstrates to be a characteristic of nature itself. Admittedly, there is something which Newtonian mechanics cannot achieve, and which explains why Newton needs god to exert an influence on the world, but Voltaire missed it. He believed that his understanding of motion – and with it the role he ascribed to god – was founded on Newton's system.[40] He did not realize that Newton's reasons for bringing god into discussion is caused by the boundaries of his own physical theory and of physics in general. (Here the point is only about physics or about that Newton called "natural philosophy").

According to Newton, god elaborates a plan for the world, sets the start-up conditions for the universe, and intervenes regularly to take corrective action. Newton's god has to place the stars and planets in their orbits, as the physical laws do not provide the initial conditions. Every now and then, god has to wind up the clock of the universe, because there are limits to the scope of the theory, and this is a flaw which would otherwise end in disaster (gravitational collapse). Finally, Newton needs god as a mediating agent because the physical law merely describes functional relations between the quantities it embraces; it is not a cause-and-effect formula. god, then, compensates for real deficiencies in Newton's physical theory.[41]

Voltaire was looking for a total explanation for the world and thought he had found it in Newtonian mechanics. He equated classical mechanics with reality, which is why the theory's limited application did not strike him as a problem. There was not enough of the physicist in him to stumble over these limits in solving physical prob-

---

38 "Si la matière gravite, comme cela est démonstré, elle ne paraît pas graviter de sa nature, ainsi qu'elle est étendue de sa nature: elle a donc reçu de DIEU la gravitation." (Voltaire, *Eléments de la philosophie de Newton*, op. cit., p. 26 [5].)

39 "Il a imprimé une loi à tous les corps, par laquelle ils tendent tous également à leur centre." (Ibid., p. 76.)

40 Cf. e. g., ibid., p. 26 [4 f.].

41 It is certainly symptomatic of Newton's inability to perceive the philosophical weight of his own accomplishment if he is obliged to call the limitations of a physical theory – determined by its status as a measuring and calculating science and, to a lesser extent, by the state of that science in his day – "god". Nevertheless, the fact remains that the tasks he entrusts to god accurately indicate the very points which, from today's perspective, stake out the limits of the physical theory "classical mechanics".

lems, and as a philosopher he was too unaware of the limits of empiricism to understand them as an epistemological necessity.

If Newton's god compensates for a deficiency in a physical theory, so Voltaire's god compensates for a flaw in Voltaire's own philosophical approach.

Metaphysics and Experimental Science in Voltaire

One of the limits to Voltaire's approach is expressed in his statement: "Metaphysics has nothing to do with geometry."[42] By "metaphysics" he means "philosophy", and by "geometry" he means "physics". He held matters of natural philosophy to be matters of physics, and these, in turn, to be matters of mathematics. However, these things can only be equated with one another if the specifics which distinguish them are not recognized. Voltaire's misreading of Newton's efforts to grasp the essence of gravitation is one indication that he saw these perspectives as interchangeable.

To be fair, it was quite common during the period to equate mathematics with mechanics, and physics with natural philosophy, but in founding classical mechanics Newton had actually given these three disciplines different roles to play. The task for the ensuing years was to comprehend these different concepts and to understand why it was so crucial that the status of experimental method, which set its stamp on the whole conceptuality of physics, be subjected to philosophical reflection.

Voltaire's view of the relationship between mathematics, physics and natural philosophy rests essentially on his failure to comprehend experimental method. We see that Voltaire holds the experimental approach in very high esteem when he writes: "True physics consists in a proper determination of all effects. It is given to us to calculate, to weigh, to measure, to observe; that is natural philosophy; almost all the rest is chimera",[43] or elsewhere: "Geometry has taught us many truths, metaphysics very few. We weigh matter, we measure it, we decompose it; and if we seek to advance one step beyond these gross operations, we find ourselves powerless, and before us an immeasurable abyss."[44] We might even be led to believe that he shares Newton's view on this, and that he has read classical mechanics appropriately.[45] He has not. He thought that measurement was simply a matter of observation, as measure and number already existed in nature. "All animals, vegetables, and minerals are arranged with due regard to weight and measure, number and motion. All is performed by springs, lever, pulleys ... ."[46] Underlying this, no doubt, is the relatively justifiable conviction that nature can be understood empirically by means of mechanics, but contemplation

---

[42] Voltaire, "Metaphysics", in *Voltaire's Philosophical Dictionary*, op. cit., vol. VII, p. 270.

[43] "La véritable physique consiste donc à bieu determinér tous les effects. Nous connaîtrous les causes premières quand donné de calculer, de peser, de mesurer, d'observer; voilà la philosophie naturelle; presque le reste est chimère." (Voltaire, "Cartésianisme", in *Dictionnaire Philosophique*, op. cit., p. 313.)

[44] Voltaire, "Matter", in *Voltaire's Philosophical Dictionary*, op. cit., vol. VII, p. 242.

[45] Cf. also Voltaire's polemical assault on Descartes, "Cartésianisme", ibid., p. 313-317.

[46] Cf. Voltaire, The Clergyman and His Soul, op. cit., p. 18; see also: Voltaire, "Les Oreilles du Comte de Chesterfield," op. cit., p. 126.

alone does not produce experimental results. It calls for complicated intellectual and physical (gegenständliche) work before it is given to us to calculate, measure, observe and hence to arrive at a geometry which can teach us many truths. Voltaire, however, does not ask what our abilities and the character of mechanics build on. He assumes that the theory is given. The only purpose of measuring and calculating, then, is evidently to find the values. Seen from this angle, there really is no difference between mathematics and physics. If one assumes that mathematically formulated physics (mechanics) need only to be taken from the peg, it is perfectly easy to adopt a rationalist standpoint, but equally an empiricist one. In both cases, the essential epistemological foundations of physics are overlooked, and inevitably speculation and metaphysics will seem to be superfluous. This always implies, not only a polemical rejection of conceptual hypostasis and dogmatism, but also the dismissal of cognitive theory and the theory of category. That is why all Newton's "metaphysical conclusions" about the nature of gravitation, the need for active principles, the atomist basis on which the unity and the lawfulness of nature are constructed, strike Voltaire as horribly obscure and best forgotten.[47]

Seen from the opposite end of the spectrum, if human thought is not in itself an object for scientific study, the epistemological foundations of physics must of necessity be ignored. It may seem strange that Voltaire pleads the claim of reason and yet refuses to call philosophy – in the sense of thinking about thoughts (Hegel) – into the witness box. In terms of Voltaire's own intellectual fabric, however, it is consistent. Philosophy's presence is not required because the methods of mathematics are equated with those of physics. Reducing Newton to a mathematician is the first step towards denying the importance of philosophy in physics.[48]

How Voltaire Links Descartes and Locke

Even if Voltaire was so blissfully unaware of the specific nature of physics, he did at least explain the great Newton's discoveries extremely well to his compatriots, and in this sense he performed a service of historical merit, embedded within the tasks of the age. However, nobody could have substantiated classical mechanics using Voltaire's understanding of philosophy or physics. Putting mathematics on a par with physics and excluding philosophy from any scientific appreciation of nature presupposes – as we have argued – that physical theory is already given. And only if that is the case can it happen that theory – in this instance mechanics – is not recognized as theory, and that it is equated with reality in a typically mechanistic way.

---

47 Ultimately Voltaire rejects metaphysics because it dares to probe the essence of god, whose tremendous power we may sense, but which remains beyond our grasp across the ravine of infinity (Cf. "Metaphysics", in *Voltaire's Philosophical Dictionary*, op. cit., vol. VII, pp.269 f.). However, precisely that is at issue: whether man is separated from infinity by anything, and if so how far. An epistemological analysis of Voltaire's position might take this as its point of departure.

48 Cf. H.-H. v. Borzeszkowski and R. Wahsner, "Über die Notwendigkeit der Philosophie für die Naturwissenschaft," op. cit.; "Noch einmal über das Bedürfnis der Naturwissenschaften nach Philosophie," op. cit.

Mechanics could not be filtered through the mechanicist lens until Newton had created it. Newton himself could not have adopted that perspective, because as the founder of mechanics he obviously could not abstract from the epistemological problems of physics. This is not to say that Newton completely recognized the epistemological status of physics, but the producer cannot regard the product as existent per se. Mechanism is based on an abstraction from the active subject, rendering the finished result absolute.

It was this absolutization which explained the direct fusion of a rationalist perspective with an empiricist one. This is especially striking in Voltaire's attitude to Descartes on the one hand and Locke on the other. The purpose, of course, was to make classical mechanics acceptable in France, where Cartesian rationalism prevailed, but without completely annihilating this rationalism, which was such an excellent basis for justifying the exclusive claims of reason.

Voltaire's attitude to Descartes was determined by this purpose. He paid tribute to Descartes for making the blind see, but marvelled at his errors for on those very grounds. "I do not think", he writes in his *Philosophical Letters*, "one can truly compare his philosophy in any way with that of Newton: the first is an experimental sketch, the second a finished masterpiece. But he who has set us on the road to truth is perhaps as worthy as he who since then has gone on to the end of it."[49]

Nevertheless, the Cartesian system proved unsuitable for the construction of an empirical mathematized natural science (and this was much due to Descartes' equating matter with extent). Yet such a science was indispensable. The existence of such a science, a theory for grasping the world scientifically, would prove that the human race really was capable of mastering nature and society through reason. Mechanics was the first – and at that time only – true science (true science in the sense of a theory which is not merely substantiated empirically but which inherently includes a strategy for linking the law with the phenomenon). Besides, the ascending bourgeoisie needed natural science as a theoretical basis for economically driven technological development. In other words, mechanics needed a justification.

Locke's sensualism came very close to the mark. Voltaire is full of praise for John Locke's philosophy.[50] Locke had shown that we have the concepts, that we have the universal from experience, and this is how he – railing against the metaphysical elements in French rationalism, and in particular the thesis of innate ideas – came to substantiate the approach adopted by mechanics. He did, however, insist that "universality belongs not to the things themselves, which are all of them particular in their existence".[51] By denying the reality of the universal, English empiricism asserts that anything declared to be reason or a law is nothing more than custom, exercise or habit. Only particular things or particular facts are real.

This, of course, fails to explain how we reach these universal concepts and how, with their help and by applying acknowledged laws or reason, we manage to find our

---

[49] Voltaire, *Philosophical Letters*, op. cit., p. 64.

[50] Cf. ibid., Letter "On Mr. Locke", pp. 52-59.

[51] J. Locke, *An Essay Concerning Human Understanding*, ed. by J. W. Yolton, London 1964, vol. II, p. 20.

way around reality. No light is shed on why what empiricism finds positive actually functions. This is why Hegel spoke of metaphysicizing empiricism.[52]

Voltaire adopted the empirical standpoint without recognizing its problems, why he assumed without more ado that it could be combined with the Cartesian method. Newton had not explicitly criticized Locke's philosophy either. Except that Newton made his standpoint productive and founded classical mechanics. Inevitably, and only to some extent consciously, he was therefore obliged to cede his patch. Locke, then, does not supply the metaphysics for Newton's theory,[53] and Hegel's assertion can be proven wrong.[54]

Reason's justified claim to be the exclusive criterion for any action could not be demonstrated from an empiricist angle. It would have been a poor argument to substantiate that claim on the basis of custom, practice or habit. In any case, it was precisely deep-entrenched custom which needed to be dislodged, and the reasonable circumstances which were "given" in England still had to be created across the Channel. There had to be some way of linking English empiricism with French rationalism, Locke with Descartes, and thus of making Newtonian mechanics acceptable to France and using it to further the cause just described. There was nothing to be gained, naturally, from agonizing over the limited application of mechanics, because limits would have been a synonym for gaps, flaws in the reason.

It was Voltaire who found the path.[55] So he took a share in that France became the focus of further development in the field of mechanics during the ensuing period. D'Alembert, Laplace and Lagrange doubtless did not derive their knowledge of Newtonian theory from Voltaire's account. But Voltaire had prepared the intellectual ground, making their work possible and providing them with a response. Emil du Bois-Reymond describes Voltaire's impact as follows: "Given the omnipotence of the court and the aristocracy, the influence of the women and the abbés, it was by no means a matter of indifference which theory they chose to embrace, and once Voltaire's *Eléments* had nudged Fontenelle's *Mondes* from the ladies' dressing-tables, Newton's victory over Descartes was as good as complete."[56]

---

52 Cf. G.W.F. Hegel, *Vorlesungen über die Geschichte der Philosophie III, Werke in 20 Bdn., auf der Grundlage der Werke von 1832-1845 neu edierte Ausgabe*, Frankfurt/Main 1986, vol. 20. pp. 209, 223.

53 Cf. ibid., p. 233.

54 Cf. H.-H. v. Borzeszkowski, "Zum Raum-Zeit-Begriff bei Locke und Newton," in *Locke-Kolloquium. Halle 1985, Kongress und Tagungsberichte der Martin-Luther-Universität Halle*, Halle 1988; R. Wahsner, "Bemerkungen zum Verhältnis von Newton und Locke," in: ibid.; R. Wahsner and H.-H. v. Borzeszkowski, "Lockes philosophischer Empirismus und Newtons empirische Wissenschaft," in *Die Wirklichkeit der Physik*, op. cit., pp. 40-47.

55 Voltaire did not overcome Cartesian dualism, but he did try to redraw the boundaries between res extensa and res cogitans – in doing so, res cogitans dwindled in the process, and it may have seemed to all appearances that the problem of duality had been reduced.

56 "Bei der Allmacht des Hofes und des Adels, dem Einfluß der Frauen und der Abbés, war es durchaus nicht gleichgültig, welcher Theorie diese huldigten; und erst als Voltaires *Élemens* Fontenelles *Mondes* von den Toiletten der Damen verdrängt hatten, konnte Newtons Sieg über Descartes für vollständig gelten." (E. du Bois-Reymond, "Voltaire als Naturforscher," in *Reden von Emil du Bois-Reymond. Erste Folge. Literatur – Philosophie – Zeitgeschichte*, Leipzig 1886, pp. 8 f.)

Voltaire used mechanics to goad the clergy and the absolutist monarchy, as a line of defence for maintaining that the world could be grasped in terms of laws (assuming a lawgiver and a world engine). Of course, he did not realize that he was re-interpreting Newton to make him palatable to the French. Drawing on English contemporaries such as Bentley, Cotes and Clarke, who had already modified Newton's standpoint for other reasons, Voltaire became the founder of French newtonianism. It was primarily in this form that Newton's ideas were digested and disseminated in the following years. Thus the prejudice was born that mechanics is mechanistic, and with it the assumption that it provided a foundation in natural science for metaphysical materialism, also known as "mechanical materialism". But this was as inaccurate as the belief that metaphysical materialism or metaphysicizing empiricism or even mechanism could be eliminated simply by founding a new theory in the natural sciences. Even today, there are voltairizations of modern science – even if they lack the Voltairean wit.

The Controversy between Voltairized Newtonianism and German Academic Philosophy

In England, the reaction to Voltaire's *Elements* was not, broadly speaking, that euphoric, although the work was quickly translated into English (in 1738) and Voltaire was made a member of the Royal Society in 1743 primarily because of this work. The prevailing opinion was that, for a Frenchman, he had not done a bad job of understanding Newton, but that England itself had not gained anything new from his work.

To begin with, reaction in Germany to Voltaire's main treatise on Newtonian physics was equally restrained.[57] Its publication was reviewed, sometimes in well-disposed vein and sometimes critically, in various relevant journals.[58] There were also reviews, however, of publications which challenged the *Eléments* along Cartesian lines, in particular Jean Banière's *Examen et réfutation des Eléments de la philosophie de*

---

[57] For more detailed accounts of the German response to Voltaire's *Eléments de la philosophie de Newton* and *La Métaphysique de Newton* see: E. L. Rathlef, "Die Geschichte des Herrn Franz Arouet von Voltaire," in Rathlef, *Geschichte Jetztlebender Gelehrten*, Siebender Theil, Zelle 1743, pp. 29-148; H. A. Korff, *Voltaire im literarischen Deutschland des XVIII. Jahrhunderts. Ein Beitrag zur Geschichte des deutschen Geistes von Gottsched bis Goethe*, Heidelberg 1917, pp. 197-235; A. Kleinert, *Die allgemeinverständlichen Physikbücher der französischen Aufklärung*, Aarau 1974, pp. 87-106; R. L. Walters and W. H. Barber, "Introduction to *Eléments de la philosophie de Newton*," in *The Complete Works of Voltaire*, ed. by W. H. Barber and U. Kölving, vol. 15, Oxford 1992, pp. 97-118; W. H. Barber, "Introduction to 'Courte réponse aux longs discours d'un docteur Allemand'," in *The Complete Works of Voltaire*, vol. 15, op. cit., pp. 751-754.

[58] Cf. *Bibliothèque germanique XLIII* (1738), 227; *Göttingische Zeitungen von Gelehrten Sachen I* (1739), 29; *Deutsche Acta Eruditorum 230* (1739), 77-102; *Nova Acta Eruditorum* (1740), 511-516.

Newton de M. de Voltaire,[59] and Regnault's *Lettre d'un physicien sur la philosophie de Neuton mise a la portée de tout le monde par Mr. de Voltaire.*[60]

Voltaire's *Metaphysik des Neuton oder Vergleichung der Meinungen des Herren von Leibnitz und des Neutons* (Newton's Metaphysics or A Comparison of the Opinions of Mssrs. Leibniz and Newton) – basically the metaphysical chapter from his *Eléments* which Voltaire fearing the censorship had originally left out – was published separately in 1740, sparking little interest in France, but in Germany it triggered a lively debate. The bone of contention – as previous reviews had hinted – was Voltaire's critique of Leibniz.[61] A voltairized Newton was juxtaposed here against a popularized version of Leibniz. Voltaire's critique deeply offended Christian Wolff's adherents, if not the entire philosophical establishment. Leibniz, it was felt, was to be dislodged in favor of a man who was respected by European scholarship as an extraordinary mathematician and physicist, but not as a great philosopher. Ludwig Martin Kahle's *Vergleichung der Leibnitzschen und Neutonischen Metaphysik wie auch verschiedener anderer philosophischer und mathematischer Lehren beyder Weltweisen angestellet und dem Herrn von Voltaire entgegen gesetzet*, published in 1741, was seen as a necessary riposte to Voltaire's insolence. Kahle found Newton's metaphysical ideas wanting and stood up for Leibniz.[62] He criticized Voltaire for calling Newton a metaphysicist, for the Englishman had merely formulated a few disparate metaphysical propositions. Voltaire could only rank him higher than Leibniz by passing over the German's great service to this branch of worldly wisdom (Weltweisheit) in silence.[63] His main purpose must surely have been "to find favor with the English".[64] He had read the German philosopher's work but fleetingly, or else forgotten the crux of it. There then followed nine chapters, reflecting the structure of Voltaire's book, in which Kahle scrutinized Voltaire's key postulations one by one and countered them in a most learned manner with all the tools of academic philosophy, examining Voltaire's statements individually from a primary perspective of metaphysical definition and formal logic founded on his own philosophical tenets, and seeking to construct evidence that his adversary has been guilty of errors of metaphysical deduction.

Nothing in the entire polemics (apart from the suggestion of a discussion about the true measure of force) related to Newton's physics, nor to the best method for measuring and calculating the motion of natural bodies; its purpose was the unconditional defence of Leibniz' philosophy, but in the wake of a previous Leibniz-Newton

---

59 Cf. *Göttingische Zeitungen von Gelehrten Sachen I* (1739), 881 f. This is the work to which Voltaire referred in his *Defence of Newtonianism.* (Cf. E. L. Rathlef, "Die Geschichte des Herrn Franz Arouet von Voltaire," op. cit., p. 114.)

60 Cf. *Göttingische Zeitungen von Gelehrten Sachen II* (1740), 393; *Nova Acta Eruditorum* (1740), 517 f.

61 Cf. *Nova Acta Eruditorum* (1742), 87-93; *Journal Littéraire d'Allemagne I* (1742), 11, 373-395; *Göttingische Zeitungen von Gelehrten Sachen II* (1740), 777-782; *III* (1741), 233; *VII* (1745), 67 f.

62 Cf. L. M. Kahle, V*ergleichung der Leibnitzschen und Neutonischen Metaphysik wie auch verschiedener anderer philosophischer und mathematischer Lehren beyder Weltweisen angestellet und dem Herrn von Voltaire entgegen gesetzet*, Göttingen 1741, Vorrede.

63 Cf. ibid., p. 115.

64 Cf. ibid., p. 29; cf. also: *Göttingische Zeitungen von Gelehrten Sachen III* (1741), Vorrede.

debate which is above all reflected in the correspondence between Leibniz and Clarke. There are no traces of any scientific progress having taken place in the thirty intervening years. Rather, this is a theologized, popularized reading of the debate, and the philosophy of Leibniz has been stripped of its flesh and blood in much the same way as the physics of Newton.

Even Christian Wolff was contemptuous of the attacks which Voltaire had launched "in the style of the feuilleton on Leibniz' profound philosophy". On one occasion he went so far as to decry "the Newtonian Philosophy, named so from simple-mindedness, deserving not even to wear the name of Physics, let alone that of all Philosophy".[65] In Wolff's opinion, it was no longer necessary to teach the "faded adulterated Freemasonry of the English".[66] He feared that "the principia of today's celebrated Englishmen" would encourage the rapid spread of materialism and scepticism, whereas his own philosophy provided the only tools to combat them.[67]

Once Voltaire was able to read Kahle's tract, translated into French in 1744, he wrote a very sarcastic letter to the author[68] and a public reply, written in the same spirit, entitled: *Courte réponse aux longs discours d'un docteur Allemand*, which ends with the words: "You wish to instruct me in the essence of things; and I teach you, sir, that neither you nor I have the honor of knowing it; I teach you that nobody

[65] Cf. Ch. Wolff, "Letter to E. Ch. von Manteuffel of 1 October 1740," in H. Wuttke, "Vorwort" to: Christian Wolffs eigene Lebensbeschreibung, in *Christian Wolff, Gesammelte Werke*, ed. by J. École, H. W. Arndt, C. A. Corr, J. E. Hofmann and M. Thomann, I. Abteilung. Deutsche Schriften, vol. 10. Biographie, Leipzig–Breslau 1739, Leipzig 1841, Halle 1755, p. 70; Ch. Wolff, "Letter to Manteuffel of 27 January 1741," in *Ch. Wolff Selbstbiographie*, ed. by H. Wuttke, Leipzig, 1841, in *Christian Wolff, Gesammelte Werke*, op. cit., I. Abteilung. Deutsche Schriften, vol. 10, op. cit.; Ch. Wolff, "Letter to J. D. Schumacher of 25 March 1741," in *Ch. Wolff, Gesammelte Werke*, op. cit., I Abteilung. Deutsche Schriften, vol. 16. Briefe von Christian Wolff aus den Jahren 1719-1753, St.Petersburg–Leipzig 1860, pp. 133 f.

[66] Cf. Ch. Wolff, "Letter to E.Ch. von Manteuffel of 14 June 1739," in *H. Ostertag, Der philosophische Gehalt des Wolff-Manteuffelschen Briefwechsels*, in *Christian Wolff, Gesammelte Werke*, op. cit., III. Abteilung, vol. 14, p. 38; also H. Wuttke, "Vorwort" to *Christian Wolffs eigene Lebensbeschreibung*, in: ibid., I. Abteilung. Deutsche Schriften, vol. 10, op. cit., p. 69, see also 70 f.

[67] "Es wäre nämlich durch die principia der heutigen berühmten Engelländer der Materialismus und Scepticismus in Italien überall gewaltig eingerißen. Man hätte sich nicht im stande gefunden aus der scholastischen Philosophie demselben zu begegnen. Daher hätte man sich mit Macht auf meine Philosophie legen müßen, weil man darinnen die Waffen gefunden, dadurch man diese Monstra bestreiten und besiegen kan. In F r a n c k r e i c h reißet der Deismus, Materialismus und Sceptocismus auch gewaltig und mehr ein, als fast zu glaubrn stehet. Und es wäre gut, wenn die vortrefflich gelehrte Marquisin gleichfals das Instrument seyn könnte, wodurch diesem Uebel vermittelst meiner Philosophie gesteuert würde." ("Letter to E.Ch. von Manteuffel of 14 August 1740," in *Christian Wolffs eigene Lebensbeschreibung*, op. cit., pp. 177 f., see also 175-184.) – For greater coverage of Wolff's attitude to Newtoniansim see: H.-H. v. Borzeszkowski and R. Wahsner, "Christian Wolff's Mechanical Philosophy: A Comparison with Isaac Newton's Mechanics," Preprint 147 of the Max Planck Institute for the History of Science, Berlin 2000 [also: Appendix to the present volume, pp. 109-134].

[68] Voltaire, "Letter to L.M. Kahle of March (?) 1744", in *Correspondence and Related Documents*, op. cit., vol. 93, pp. 78 f. (D 2945).

has lived on this earth until today, nor ever will, who knows what matter is, what the principle of life, of sentiment, or what the human soul!"[69]

During the entire debate, Newton's physics, its epistemological foundations and consequences, had – as we have seen – no part to play (not even the "Regulae Philosophandi"). Discussion revolved around individual passages in Newton's writing which, in the main, had some bearing on religious philosophy, albeit in Voltaire's rendering. No attempt was made to verify whether Newton had actually written anything of the kind himself.[70]

In Germany, there was a widespread failure to recognize Voltaire's philosophical agenda or, indeed, to acknowledge him as a philosopher at all. No doubt, the accusation that he had not understood Leibniz is perfectly legitimate (just as he did not do justice to Newton either). However, one must add that the (pre-Kantian) German philosophers did not understand Newton either; or, to be more precise, they failed to see the *philosophical* relevance of Newtonian physics.[71] (This, of course, does not detract from the merits of the German Enlightenment or rationalist philosophy in Germany, and is not intended to reduce either to this critical factor.)

Voltaire was criticizing the philosophy of Leibniz and Wolff insofar as it had become a Gospel, and the dogmatic self-evidence of arguing on the basis of this system. He was attacking a philosophical system insofar as it was a system, and attacking the metaphysics in rationalist guise. (Philosophy could not regain its status as a system until its underwent a synthesis with empiricism.)

Even the announcement of *Newton's Metaphysics* just quoted from the *Göttingische Zeitung*, which is otherwise not well disposed towards Voltaire, unintentionally confirmed this by quoting Voltaire's own objectives thus: "In his comparison of the teachings of Newton and Leibnitz, he had merely demonstrated that Newton did not establish any system because he was able to doubt, and that in Germany a hue and cry was raised by such as said they doubted not, but knew all things for certain."[72] The real motive for the intolerant tone of the debate was resistance to scepticism, deism and materialism. After all, Voltaire had aroused German suspicions some years

---

69 "Eh bien; vous nous apprenez donc, qu'il y a des essences; et moi je vous apprends que ni vous ni moi n'avons l'honneur de les connaître; je vous appends, que jamois homme sur la Terre n'a su et ne saura ce que c'est que la matière; ce que c'est que le principe de la vie et du sentiment; ce que c'est que l'âme humaine; s'il y a des âmes dont la nature soit seulement de sentir sans raisonner, ou de raisonner en ne sentant point, ou de ne faire ni l'un ni l'autre." (Voltaire, "Courte réponse aux longs discours d'un docteur Allemand," in *The Complete Works of Voltaire*, op. cit., vol. 15, pp. 756-762.)

70 For example, the reviewer who penned the detailed announcement of Voltaire's work [*Göttingische Zeitungen von Gelehrten Sachen II* (1740), 777-782] confused the theme and chapter structure of this with the theme and chapter structure of a supposed work by Newton which in fact never existed.

71 A closer examination of this will be found in the chapter "Die Mechnisierung der Mechanik" in the volume quoted in Fn. 3 on the third place.

72 ["Er habe in seiner Vergleichung der Lehren von Newton und Leibnitz nur gewiesen, daß Newton kein Systema angenommen, weil er zu zweifeln gewußt. Und darüber habe sich in Teutschland ein Geschrey erhoben von Leuten, die da sagen, sie zweifelten nicht; sie wüßten alles gewiß."] *Göttingische Zeitungen von Gelehrten Sachen VII* (1745), 67.

previously with his *Philosophical Letters* by gnawing at the dualism of soul and body, and thereby challenging immortality. Voltaire had stated in his letters that, bearing in mind god's omnipotence, he could imagine souls thinking or, indeed, matter thinking. It was the latter which was held to be the most dangerous error. The heftiest criticism concerned the "fluid and pleasant style of writing" in which Voltaire cloaked his errors. A treatise of suitable rigor was penned to condemn it.[73]

One might wonder why, given that Voltaire's *Metaphysics of Newton* (or so it was felt) was so unable to capture the depths of Leibniz' metaphysical ideas, people did not simply ignore it, or perhaps choose the option of taking the author's arguments to pieces with calm objectivity. Presumably the reason, apart from an abhorrence of scepticism, was the memory of a previous debate about Leibniz and Newton which had begun in the late 17th century, culminating in an exchange of letters between Leibniz and Clarke.[74]

Although on the face of it this was a debate about god and the quality of his creation, the scientific core of this dispute was to determine the concept of an empirical mathematized natural science, in contradistinction to both the concept of philosophy and that of purely empirical natural research, or the pure description of nature.[75] Newton (in the person of Clarke) adopted the stance of empirical mathematized natural science, while Leibniz defended the tasks of philosophy. Neither were aware the core of the dispute (how could they?), with the result that they tended to talk past each other. However, there were tentative indications of the distinction in as much as Newton in particular stresses the need to differentiate between "experimental philosophy" and "speculative philosophy".[76] But it was not until Kant that the distinction between philosophy and natural science was elevated into philosophical consciousness.

---

[73] In 1740 Johann Gustav Reinbeck published this under the (German) title *Philosophical Thoughts on the reasoning soul and the immortality thereof accompanied by some remarks on a French work which claims that matter thinks.*

[74] Wolff's review of the original English edition of this correspondence appeared in *Acta Eruditorum* in October 1717; three or four months later Bernard's followed in *Nouvelles de la République des Lettres*; the German edition, published in autumn 1720, contained a foreword by Christian Wolff, not surprisingly defending the Leibniz position; this edition was reviewed the same year in *Acta Eruditorum*, and the following year in *Historie der Gelehrsamkeit Unserer Zeiten.* It was in 1721, too, that the *Deutsche Acta Eruditorum* announced the French edition of the correspondence between Leibniz and Clarke, coupling this with a comment. This is merely a selection.

[75] Cf. on this dispute about Newton and Leibniz: E. Cassirer, "Newton and Leibniz," *The Philosophical Review II* (1943), No. 310; J.O. Fleckenstein, *Der Prioritätsstreit zwischen Leibniz und Newton*, Basel–Stuttgart 1956; A. Koyré and I. B. Cohen, "Newton and the Leibniz-Clarke correspondence with notes on Newton, Conti and Des Maizeaux," *Archives internationales d'histoire des sciences 15* (1962), 63-126; H.-H. v. Borzeszkowski and R. Wahsner, "Die Metaphysizierung der Physik," in *Weltanschauung und Denkweise*, ed. by H. Horstmann, Berlin 1981, pp. 149-161; *Physikalischer Dualismus und dialektischer Widerspruch. Studien zum physikalischen Bewegungsbegriff*, Darmstadt 1989, pp. 39-49; Introduction and Notes to the German edition of Voltaires's *Elemente der Philosophie Newtons*, op. cit.; V. Schüller, Appendix of the editor of: *Der Leibniz-Clarke-Briefwechsel*, Berlin 1989 (this volume also contains many other letters and documents pertinent to the correspondence).

[76] See e.g. the documents in the appendix to the German edition of the Leibniz-Clarke correspondence quoted above, esp. pp. 182-189, 190-194, 198-202, 210, 222-237, 239, 242, 252.

The Concept of Mechanics adopted by German Idealism

From the outset, Kant was concerned to adapt metaphysics (to him this meant the philosophy of Leibniz and Wolff) to the new science (typically expressed in Newtonian mechanics), or at least to establish a consistent link between them. Even his earliest works, his "naturalist" period, betray this orientation.[77]

The way in which the French Enlightenment had linked English newtonianism with Cartesian rationalism led, on the one hand, to a brusque rejection of metaphysics and to a mechanization of mechanics, but on the other it had opened up traditional metaphysics to questions which had hitherto been the prerogative of empiricism. Now Kant restructured metaphysics to enable it to tackle this challenge. Although he was not the only person involved in restructuring metaphysics sufficiently to found epistemologically the new physics,[78] he took the decisive step.[79] He was able to do this for, on the ons hand, Leibniz' philosophy had absorbed a critique of both Descartes and Locke, and, on the other hand, Kant was never in real danger of succumbing to empirical extremities. The empiricist roots of his system did not derive from any current of empiricist philosophy, but from the concept of experience found in mathematical physics. His teacher Knutzen, a disciple of Wolff and also a keen champion of Newton, believed that the developments in natural science were extremely important to philosophy, and it was he who introduced Kant to Newton's work. From the outset Kant was enthralled by the challenge of resolving inevitable contradictions in the philosophical standpoint of his revered teacher. The way forward was evidently to link Leibniz and Newton on a consistent basis. The key to this was Kant's recognition of the basic difference between the sensuous world and the intelligible world.

Kant's "critical philosophy" was founded on a metaphysical analysis of Newton's method and a restructuring of Leibnizian metaphysics using the insights derived from that analysis. It can be seen as a *philosophical* comprehending of Newton's theory or as a recognition of the philosophical consequences of modern-age *(neuzeitlich)* natural science. This *Kritizismus* is a synthesis between rationalist German philosophy and empiricist English philosophy caught in the prism of the French Enlightenment. If only by means of methodological regularization, Kant synthesized Leibniz and Newton, whose conceptions had been the subject of such vehement and controversial debate in the German-speaking territories since the end of the 17th century. The objective

---

[77] A convincing case for this is made by E. Cassirer in *Kants Leben und Lehre*, Berlin 1918, especially in Chapter 2, sections 1 and 2.

[78] Even Wolff lamented at the miserable condition of metaphysics. Attempts were made to improve it by supplying philosophy with a mathematical input (cf. "Ch. Wolff's prologue for the German edition of the Leibniz-Clarke correspondence," in *Der Leibniz-Clarke-Briefwechsel*, op. cit.) or by applying the methods of physics to philosophy (cf. G. Sharpe, "Eine Verteidigung von Dr. Clarkes fünfter Erwiderung an Mr. Leibniz – eine Antwort and Mr. Thümmig," in: ibid.).

[79] For a closer study see: R. Wahsner, "Die Kantsche Synthese von Leibniz und Newton und deren Konsequenzen für den Mechanik-Begriff des deutschen Idealismus," in *Konferenzakten des 9. Internationalen Kant-Kongresses, Berlin, 26. 3.- 31. 3. 2000* (in print) [also: Preprint 111 of the Max Planck Institute for the History of Science, Berlin 1999].

core of this debate had, as we have already seen, been a clash between the epistemological status of modern-age (*neuzeitlich*) empirical mathematized natural science, and an assertion of the need for philosophy.[80] Kant's epistemological turn pointed the way to a solution, and one which demonstrated that both disciplines were necessary (and possible) by substantiating the limits to each and the essential difference between them. In this way, Kant perfected and corrected the desire of the (English and French) Enlightenment to understand philosophy as a *science* of experience.

Kant's solution does, however, retain the vestige of an Enlightenment flaw, a vestige which finds expression in the mechanistic concept of mechanics which Kant (like Hegel) was not fully able to overcome. We see this in Kant's conflation of metaphysical (= Leibniz) and physical (= Newton) tasks in his works of natural philosophy,[81] a symptom that he has not fully completed a *philosophical synthesis* of Leibniz and Newton. This vestige is presumably rooted in the great significance which the category *relationship (function, order)* has for the principle of modern-age (*neuzeitlich*) thought (philosophically founded by Kant's transcendental method), although the conceptual reconstitution of this category is still inadequate.[82] Rather than thinking of gravitation (like all dynamic forces) as a one-to-another, Kant mainly conceives it as two (separate) opposite impacts.[83] This only partly surmounts the mechanistic attempt (to which even Euler was not immune) to take something which can only be constituted as a relationship and reduce it to the property of a body. Technically, this flaw makes its impact felt when Kant (and later Hegel) overlooks the fundamental epistemological significance of the second derivatives in Newton's theory, including the fact that an essential of motion, of a behavior, namely acceleration, can be conceived as an object. This is coupled with an underestimation of mathematics as the functional language of physics, a tendency to substitute illustrative models for system-based theoretical concepts (to make something plausible). The mechanistic vestige is reflected by Kant in the fact that he does not fully cease to equate metaphysical foundation with theoretical physics – often in violation of his own metaphysical principle. Leibniz equated them almost totally, except that his concept of physics was not mechanistic: instead it was highly logicistical and mathematicistical.

To the extent that Kant's concept of mechanics contains mechanistic remnants, the synthesis between Newton and Leibniz cannot be perfected, and that means that the epistemological status of an measuring and calculating science cannot be sufficiently adequately established by contradistinction and in relationship to philosophy. It is understandable that Kant did not fully overcome mechanism. In "adapting" the old

---

80 Cf. H.-H. v. Borzeszkowski and R. Wahsner, *Physikalischer Dualismus und dialektischer Widerspruch*, op. cit., pp. 39-49.

81 Cf. R. Wahsner, *Das Aktive und das Passive*, op. cit.

82 For more detail: R. Wahsner, "Die fehlende Kategorie. Das Prinzip der kollektiven Einheit und der philosophische Systembegriff," in *Wiener Jahrbuch für Philosophie* XXXI/1999, S. 43-60.

83 Cf. Hegel's critique of this: G.W.F. Hegel, *Wissenschaft der Logik. Erster Teil*, in *Werke*, op. cit., vol. 5, pp. 200-208; *Enzyklopädie der philosophischen Wissenschaften im Grundrisse (1830). Zweiter Teil. Die Naturphilosophie, mit den mündlichen Zusätzen*, in *Werke*, op. cit., vol. 9, p. 61 (§ 262 A).

metaphysics – with Leibniz' philosophy as its culmination – to the lens by which mechanics derives its insights, he was observing the latter through the spectacles of philosophy, and initially he assumed that what his philosophical environment told him was mechanics really was mechanics. The conceptual innovations contained in effect in Newton's theory, and in particular the concept of gravitation, had been filtered through the old philosophies. By developing his system, Kant gained a deeper understanding of the metaphysical consequences of Newton's method, but his picture of mechanics had hardly changed.

Philosophy did not accomplish an appropriate reading of mechanics until it was possible to compare different yet valid physical theories with one another, and physical theories with biological, chemical and other theories, which revealed what was actually typical of natural science as such and what was specific to a particular physical theory or specialist field. As we know, a situation of this kind developed in the latter half of the 19th century, once the era of Classical German philosophy was over.

Christian Wolff's Mechanical Philosophy: A Comparison with Isaac Newton's Mechanics

Horst-Heino v. Borzeszkowski and Renate Wahsner

Ever since Newtonian mechanics was born, some have felt the need to wonder whether Newton's road to a physical mechanics was the only road possible or whether there were realistic alternatives, even in his day, while others have been prompted to object to Newton's approach, equating its faultless physical substantiation with its philosophical comprehension (or rather, taking it for granted that the two things must surely coincide). By describing Wolff's conception of mechanics and comparing it with Newton's, we hope to cast some light upon some of the contentious issues in the Newton-Wolff debate which took place in the 18th century and to identify the consequences of that dispute.

The Wolff-Keill Debate

The questions which this comparison seeks to address are illustrated by the controversy which unfolded between Christian Wolff and John Keill, a British Newtonian, following the publication in 1709 of Wolff's *Aerometrie.*[1] Keill, a devoted champion of Newton's method and his concept of physics,[2] argues here in the manner customary to physics as developed from Galileo to Newton, although he makes mistakes over certain physical details. Had Wolff picked these up, he would have made light work of Keill. However, he chose a line of objection which demonstrates that he had no instinct at all for this type of natural research, that he was not even capable of following *Keill's argument,* or perhaps that he had quite different objectives in mind.

Keill wrote to Wolff after reading his *Aerometrie*, published in Leipzig in 1709, in which Wolff maintained that Keill was guilty of erroneous deductions.[3] Wolff accepted Keill's view that the "gravities of bodies are to one another as the quantity (*quantita-*

---

[1] See: "Zwey zwischen Hrn. Johann Keil Lehrern der freyen Künste in dem Christ-Hause auf der Universität zu Oxfort, und Hrn. Wolfen gewechselte Schreiben von dem leeren Raum," in *Christian Wolff, Gesammelte Werke*, ed. by J. École, H. W. Arndt, C. A. Corr, J. E. Hofmann and M. Thomann, Hildesheim–Zürich–NewYork 1962 ff., I. Abteilung. Deutsche Schriften, vol. 21.1. Gesammelte kleine philosophische Schriften I, Hildesheim–NewYork 1981, pp. 257-274.

[2] As to Keill, for further detalis see: F. Rosenberger, *Isaac Newton und seine physikalischen Prinzipien. Ein Hauptstück aus der Entwicklungsgeschichte der modernen Physik*, Leipzig 1895, pp. 288, 344-352, 468, 495-497, 500, 503, 513. Therein the correspondence between Keill and Wolff cited here is mentioned and Wolff's attitude to Newton's physics is addressed, too (espesially see: pp. 224, 233 f., 330 f.). For an interesting description of the role played by Keill in the propagation of newtonianism see: Robert E. Schofield, *Mechanism and Materialism. British Natural Philosophy in An Age of Reason*, Princeton–NewJersey 1970, especially pp. 19-62, 80, 134-156.

[3] Wolff was referring to Keill's *Introductio ad veram Physicam*, Oxford 1702.

tes) of their matter", but warned that Keill was "drawing a false conclusion" in thinking that the existence or "reality" of empty space could be deduced from this proposition.[4]

Keill had drawn this conclusion in his *Introductio* because bodies of the same size have different weights, and therefore can contain different amounts of matter. It followed, in his view, that there might be "more holes or more spaces which are entirely empty" in lighter bodies than in heavier ones.[5] Wolff objected that all bodies "swim in a very thin fluid essence which permeates their holes"[6]. Although this fluid essence did not increase the weight of the body, it nevertheless left no holes empty.

Wolff draws here on Archimedes for support, and so Keill and others can be excused for believing that the debate was a strictly physical one. Keill, therefore, tries to demonstrate in a letter to Wolff that even if bodies are suspended in a fluid the ratio between the quantity of matter of two bodies will be equal to the ratio between their weights. This enables him to repeat his earlier conclusion that bodies contain empty spaces.

To this end, Keill proves two "theorems", where the first is derived from relationships which he then establishes in the second. To prove the second, he applies the Cartesian measure for the quantity of motion, given by the product *quantity of matter* × *velocity*. Keill then proceeds to argue as follows: The ratio between the quantity of motion of two bodies which are sinking in a fluid[7] is equal to the ratio between the actual weights or gravities of the bodies in the fluid. However, due to Archimedean buoyancy, actual weight in a fluid is reduced by the weight of the amount of liquid displaced by the body (as opposed to its proper weight or inherent "heaviness"[8]). As the velocities, assuming they are the same, are deducted proportionally from the ratio between the quantities of motion, the ratio between the quantities of matter is equal to the ratio between the actual weight of the bodies in the fluid.

Wishing to draw in his line of argument on what he then proceeds to demonstrate in his second theorem, Keill attempts in his first theorem to substantiate the assumption he makes in his second that bodies in a fluid sink at the same velocity. To do this he leans on Newton and Huygens who, in his opinion, "have shown from much experience that, if we disregard the resistance of the air, all bodies describe the same space as they fall".[9] Keill feels that this assumption equips him to argue the same for bodies

---

[4] Cf. J. Keill to Ch. Wolff, quoted in the above-mentioned correspondence, op. cit., p. 258 f. (Quotations from Keill have been translated for the purpose of this article from the German edition in fn. 1.)

[5] Cf. ibid., p. 261.

[6] "Denn da alle Cörper in einem sehr dünnen flüßigen Wesen schwimmen, welches durch ihre Löcher hindurch gehet ..." (Ibid., p. 261 f.)

[7] The German translation of Keill's letter describes this as a "fluid essence".

[8] What Keill means by "proper weight" is the weight measured on scales.

[9] "Da aber die vortrefflichen Männer Newton und Hugens aus vielfältiger Erfahrung bekräfftiget haben, daß alle Cörper, wenn man von dem Widerstand der Lufft abgehet, einen gleichen Raum im fallen beschreiben: so müssen notwendig die Gewichte der Cörper sich gegen einander verhalten, wie die Grösen ihrer Materie, es mag nun der Raum, darinnen die Cörper fallen, ganz leer oder mit

sinking in liquid and to conclude that the weights of the bodies are to one another as the quantities of their matter, regardless of whether the space in which they fall is completely empty or filled with fluid.[10]

From the standpoint of physics, there is much to criticize about the way Keill develops his disquisition, for example, his assumption that the quantity of motion (as the product *quantity of matter* × *velocity*) is proportional to weight. This "Aristotelian law" is not found in Newtonian physics (nor in its post-Newtonian evolution).[11] Another assumption which has to be challenged is the idea that bodies sink at equal speed in fluid. It is risky enough from the logical perspective to found this assumption on the law of falling bodies, according to which bodies fall at equal speeds in a vacuum. It also leads to utterly erroneous physical conclusions, because bodies do not fall at equal speeds in fluid. They only do this in an empty space.[12]

Again, it would have been easier to show by purely physical arguments that Keill's case for the existence of empty space within a body was fallacious. In his answer to Keill's letter, however, Wolff passes over his line of argument in total silence and (implicitly accepting Keill's physical Ansätze and calculations) retorts instead that all this fails to prove the existence of vacuums, because although bodies contain pores of various quantities and dimensions, they are not in the least empty, but hold "alien matter" (*fremde Materie*) which itself has no heaviness.[13] Weight is determined merely by "peculiar matter" (*eigentümliche Materie*). If two bodies of the same size weigh differently, this is due in Wolff's eyes to the fact that the lighter one has less peculiar matter and more pores than the heavier one, however, these pores are full of alien matter. The ratio between the weights proposed by Keill, he argues, is no measure for the ratio between matter and vacuum in a body, but for the ratio between matter peculiar and alien to it.

This dispute is quite evidently rooted in Wolff's concepts of matter and body, motion and force, and we shall therefore scrutinize these more closely.

---

einer sehr dünnen Materie erfüllet seyn." (J. Keill to Ch. Wolff, quoted in the above-mentioned correspondence, op. cit., p. 263.)

[10] Cf. ibid., p. 263 f.

[11] It can be said, rather, that Newton's equations of motion, when friction is taken into account, permit the conclusion that velocity is proportional to weight when friction is significant. But the factor of proportionality is given by the friction coefficient, not the quantity of matter. Besides, Keill (echoing the wording of Wolff's objection) speaks of a very thin fluid essence or very thin matter, in other words, a medium in which there would be very low friction. (Cf. J. Keill to Ch. Wolff, quoted in the above-mentioned correspondence, op. cit., p. 264.) As a result, Keill bases his reasoning on an "Aristotelian law" rather than on Newtonian physics.

[12] Presumably Keill adopts Wolff's thesis about the thin fluid in order to cover up the difference between empty space and space filled with fluid. Wolff's point, however, is that empty space is not at all the same as space filled with anything, even if it is very thin fluid.

[13] Cf. footnotes 25-29, 64.

Matter, Body and Motion, Heaviness and Impact – Wolff's Interpretation

Wolff explains his own method as follows: First he considered the essence and nature of bodies and elucidated the properties which derive from these. He outlined how subtly nature divides matter and why geometry is unable to demonstrate that matter is divisible; that there is no empty space between the subtle parts of matter and that all matter is permanently in motion, which is why bodies have a variable force which we have come to call their nature. Developing these ideas further, he described how two kinds of matter are present in every body – peculiar matter and alien matter – and that there are two kinds of peculiar matter, one permanent (*beständige Materie)* and one variable (*veränderliche Materie*), and that any change to the permanent matter causes a change in essence, whereas any change to the variable or alien matter merely causes a change of state.[14] He then devoted some attention to which differences in physical things derive from permanent matter and which from variable matter. The former category concerns density, hardness of different kinds, roughness, fragility, resilience etc., while among the latter he ranks swelling, viscosity, softness, warmth, coldness, heaviness and the body's expanding force. He felt it was important, not only to explain the properties, but also to account for how they change.[15] Wolff moves on from this general consideration of bodies to a phenomenological consideration of world bodies, concluding: "In composing the universe from these world bodies, I assume nothing but what accords with observations. And thus emerges the Copernican structure of the universe, which all who understand astronomy now profess to endorse."[16] – A astonishing result, considering the complicated struggle it took to establish the Copernican world view.

Wolff devoted an analytical work to the concept of bodies, summarizing as follows: From the phenomena, I have derived that the concept of body implies extension, idle force (*faule Kraft*) and motive force (*bewegende Kraft*). The body together contains these, and this must be conceded in order to provide sufficient reason for the phenomena. I have shown, moreover, that the concepts of dimension and idle and motive force are not primary, but can be broken down into even simpler parts. The primary origins, the elements of bodies are simple things existing in themselves (*substantiae*) in which, as they are the origin of phenomena, sufficient reasons must be found as to

---

[14] This determination is not entirely unambivalent, as he concedes that under certain circumstances a change in essence may result from a change to the alien matter. (Cf. Ch. Wolff, *Vernünfftige Gedancken von der Natur und ihren Würckungen*, in *Werke*, op. cit., vol. 6 *[Deutsche Physik]*, Hildesheim–NewYork 1981, p. 34.)

[15] Cf. Ch. Wolff, *Ausführliche Nachricht von seinen eigenen Schriften, die er in deutscher Sprache von den verschiedenen Theilen der Welt-Weißheit herausgegeben*, Frankfurt/Main 1733, in *Werke*, op. cit., vol. 9, edited and with an introduction by H. W. Arndt, Hildesheim–Zürich–NewYork 1996, p. 499 f.

[16] "Indem ich das Welt-Gebäude aus den Welt-Cörpern zusammen setze, nehme ich nichts an, als was den Observationen gemäß ist. Und so kommet dann der Copernicanische Welt-Bau heraus, dem heute zu Tage alle beypflichten, welche die Astronomie verstehen." (Ibid., p. 501.)

why the bodies have extension and why they contain within them a motive force. This develops continuously.[17]

Accordingly, Wolff begins by analyzing *body* as a concept.[18] He asks first what aspects of a body must be distinguished in order to find the true reason for everything which belongs to it. Every body, tradition maintains, has extension and can, therefore, be divided into many parts. In addition, however (and this is crucial to his case against Descartes), every body has a force for withstanding motion, so that it cannot be set in motion before its resistance is broken. "What gives a body extension and its resistant force" – now the metaphysical thought unfolds – "is known as matter". Yet matter is not an untidy lump, "that is, no heap of similar parts thrown together, distinguishable by nothing but size", in other words, it does not consist of atoms, qualitativeless atoms, but everything about it is specially distinct from everything else.[19]

Wolff's concept of body implies Wolff's law of inertia, which says: "As matter has a force with which to resist motion, a body cannot move itself, and so it must have a cause from outside if it is to move. ... Just as nothing can occur without sufficient reason for it to be rather than not to be, so the motion of a body cannot cease once it has started, unless a cause be encountered from outside for this to happen. And in the same way it is evident that it must continue to move constantly towards a region, if nothing changes its direction. ... In this way any body will persist undisturbed either at rest or in motion, and will preserve the same direction, until it be placed into a different state by another: all of which we can also observe in our experience, and those who study the rules of motion accept this as an unchanging law of nature."[20] Wolff refers explicitly at this point to Newton's *Principia*.[21] All these things "we can also observe in our experience" may well be an allusion to the painstakingly acquired insights (and laws) of Newtonian mechanics, although a precise comparison reveals very different determinations of the inertial principle.

---

[17] Cf. Ch. Wolff, "Von dem Begriff des Cörpers," in *Werke*, op. cit., vol. 21.1. Gesammelte kleine philosophische Schriften I, op. cit., pp. 190-256.

[18] On Kant's subsequent critique of this method cf. R. Wahsner, "Die Kantsche Synthese von Leibniz und Newton und deren Konsequenzen für den Mechanik-Begriff des deutschen Idealismus," in *Konferenzakten des 9. Internationalen Kant-Kongresses, Berlin, 26. 3.- 31. 3. 2000* (in print) [also: Preprint 111 of the Max Planck Institute for the History of Science, Berlin 1999].

[19] Cf. Ch. Wolff, *Vernünfftige Gedancken Von Gott, Der Welt und der Seele des Menschen, Auch allen Dingen überhaupt*, in *Werke*, op. cit., vol. 2.2 *[Deutsche Metaphysik]*, ed. by C. A. Corr, Hildesheim–Zürich–NewYork 1997, pp. 375-377.

[20] "Da die Materie eine Kraft hat der Bewegung zu widerstehen, so kan kein Cörper sich selbst bewegen, und demnach muß es von aussen eine Ursache haben, wenn er sich bewegen soll. ... Wiederum weil nichts geschehen kan, was nicht seinen zureichenden Grund hat, warum es vielmehr ist als nicht ist, so kan auch die Bewegung eines Cörpers nicht wieder aufhören, darein er einmal gesetzet worden, wenn nicht von aussen eine Ursache anzutreffen, warum es geschiehet. Und auf gleiche Weise erhellet, daß er fortfahren muß sich beständig nach einer Gegend zu bewegen, wenn nichts seine Richtung ändert. ... Auf solche Weise verharret ein jeder Cörper unverrückt entweder in seiner Ruhe, oder seiner Bewegung, und behält einerley Richtung, bis er durch einen andern in einen anderen Zustand gesetzet wird: welches man auch alles in der Erfahrung also findet, und von denen, welche die Regeln der Bewegung untersuchet, als ein unveränderliches Gesetze der Natur angenommen wird." (Ibid., p. 377 f.)

[21] He quotes p. 12 edit. sec.

According to Wolff, the essence of a body depends on its composition, although he comments that natural science cannot delve as deep as the elements, but stops short at conglomerates, that is, corporal parts. Wolff hopes to show that every little particle of matter really is divided into an infinite number of parts, but that these are unlike each other in size, shape and every other respect. In other words, they are not atoms. However, he ponders, there must be some way to divide matter, and so he develops a metaphysical argument that this is achieved by motion, which is why all matter in nature must be permanently in motion: "and so we must recognize above all else that there can be no little spaces empty of all matter between the parts of matter which are found in a body".[22] (Hence his vehement dispute with Keill.)

This demand may seem somewhat arbitrary in itself, but it is consistent with Wolff's system once we realize that Wolff's idea of motion, or force, was quite simply impact. Wolff's metaphysics teaches that if one body receives an impact from another, it pushes back with its resistant force, thereby resisting motion. It is clear from this, he believes, that the magnitude of the body's resisting force will be determined by the magnitude of the impact, and that far from having definite limits of its own, it is the impact which sets the limits, the impact in which we encounter the reason for its change. A motive force, Wolff declares, can increase and decrease in a body, for if a body is moving at a certain velocity, a further impact can give it greater velocity, and thus the motive force can be increased (or, equally, decreased). By this token, a body has a greater force if it is moving fast because, Wolff explains, a force cannot be changed except by degree, and so motive force is changed by the degree of velocity in the same way that the extension of matter is changed by its shape. Finally, he concludes: "We find in our experience that no change can occur to the velocity of a body except by impact. ... Similarly, experience shows that no body can be set in motion from rest unless it receives an impact from another, which is in motion. And in this manner motion can come only of previous motion."[23]

Gravitation is conceived in complete harmony with this concept of force and motion. Wolff understands it simply as heaviness. He offers an explanation: "Heaviness (*gravitas*) in mechanics is a force by which a body is driven towards the mid-point of the earth. Or: Heaviness consists in an effort to move towards the mid-point of the earth."[24] This concept of heaviness corresponds to Wolff's position in the dispute

---

22 "Damit wir die Nothwendigkeit sehen, warum alle Materie beständig in Bewegung seyn müsse; so müssen wir für allen Dingen erkennen, daß zwischen den Theilen der Materie, die sich in einem Cörper befinden, keine Räumlein seyn können, die von aller Materie leer sind" (Ch. Wolff, *Vernünfftige Gedancken von der Natur und ihren Würckungen [Deutsche Physik]*, op. cit., p. 20.)

23 "Wir finden in der Erfahrung, daß keine Veränderung in der Geschwindigkeit eines Cörpers geschehen kan, als durch den Stoß. ... Eben so zeiget die Erfahrung, daß kein Cörper aus der Ruhe in Bewegung gesetzet wird, als durch den Stoß eines andern, der in Bewegung ist. Und auf solche Weise kommet keine Bewegung als aus einer vorhergehenden Bewegung." (Ch. Wolff, *Vernünfftige Gedancken Von Gott, ... [Deutsche Metaphysik]*, op. cit., pp. 405-409.)

24 "Schwere, gravitas, ist in der Mechanick eine Krafft, durch welche der Cörper gegen den Mittel-Punct der Erde getrieben wird. Oder: Die Schwere bestehet in einer Bemühung, sich gegen den Mittel-Punct der Erde zu bewegen." (H. A. Meissner, *Philosophisches Lexicon aus Christian Wolffs sämtlichen deutschen Schriften*, Bayreuth and Hof 1737, p. 537.)

about gravitation as an inherent property of matter – although it is still unclear at this point whether his concept of heaviness derives from this position or whether his position derives from the concept. In his *Deutsche Physik* he goes into greater depth: "Among the properties of bodies which derive from alien matter I include heaviness … . I am well aware that there are many in England today who say that heaviness is peculiar to all matter and that it is therefore proportional to the matter contained in any body, and has no mechanical cause which would explain it. … They assume the former in order to be able to demonstrate that there are little spaces which are empty of all matter and that in consequence there are also indivisible parts of matter which have a necessary shape and dimension, and also a perfect hardness. … [That is why] we shall have all the more cause to show that not all matter is heavy and that heaviness does have its mechanical cause, that is, draws its origin from its motion, according to the regular rules of the latter, observed in nature in the motion of other bodies."[25] He now paints consequences for the structure of the world: Heaviness is an effort to move towards the centre of the earth. Suffice it to understand that, as heavy bodies move towards the earth by means of their heaviness, this heaviness can be regarded as a motive force. As we see that matter, which is heavy, does not move in a single direction but in many, we also see that the direction is not necessary, and so not derived from the essence of matter. It follows that heaviness is not peculiar to matter and, by consequence, that matter need not necessarily be heavy.[26] Trying to avoid this consequence by referring to god simply means declaring heaviness as an occult quality, i.e. something which has no cause. Wolff draws his conclusion from this line of argument: because there is no sufficient reason in heavy bodies for their heaviness, and yet the principle of sufficient reason dictates that there has to be one, there must be something apart from heavy matter where this can reside. The reason, he continues, must derive from the motion of some other matter which drives heavy bodies [heavy per se?] towards the centre of the earth. "And so there is a special matter in nature which is permanently in motion, which by means of this motion makes bodies heavy", i.e. drives them towards the earth's centre. "And this matter is what we call heavy-making matter." Experiments, claims Wolff, had shown that a heavy body constantly changes its motion as it begins to fall, moving continually faster and faster. As velocity, in his view, can only be changed as the result of a new impact, this heavy-making matter must be

---

25 "Unter die Eigenschafften der Cörper, welche von einer fremden Materie herrühren, rechne ich auch die Schweere … Ich weiß wohl, daß heute zu Tage verschiedene in Engelland vorgeben, die Schweere sey aller Materie eigenthümlich und daher in einem jeden Cörper der in ihm enthaltenen Materie proportional, und habe keine mechanische Ursache, daraus sie sich erklären lasse. … das erstere nehmen sie deswegen an, damit sie in dem Stande sind zuerweisen, es gebe Räumlein, die von aller Materie leer sind und folgends auch untheilbare Theile der Materie, die eine nothwendige Figur und Grösse, auch dabey eine vollkommene Härte hätten. … (Deshalb) werden wir um so vielmehr Ursache haben zuzeigen, daß nicht alle Materie schweer sey und daß die Schweere allerdings ihre mechanische Ursache habe, das ist, aus der Bewegung ihren Ursprung nehme, nach den *ordentlichen Regeln derselben, die in Bewegung anderer Cörper von der Natur beobachtet werden."* (Ch. Wolff, *Vernünfftige Gedancken von der Natur und ihren Würckungen [Deutsche Physik]*, op. cit., pp. 116 f.)

26 Cf. ibid., pp. 117-119.

constantly pushing the heavy body, and at the same time it must be spread throughout all the space which the earth has filled with air.[27] As heaviness is not determined by the external surface of a body, the matter which makes it heavy obviously cannot simply be pushing its surface, and so it is not possible that heaviness only exerts its influence on the body from the outside; evidently it must be able to penetrate freely into the interstices of even the densest bodies.[28] This latter circumstance means that parts of the heavy-making matter must constitute something continuous, and that as a result of this they cannot really be divided. However, they *are* divided and so they cannot constitute this continuous thing, but must have interstices which are filled by another matter with different forms of motion. Apart from heavy-making matter, then, there must also be other, subtler types of matter which also have no heaviness.[29] And because the heavy-making matter does not always move together with a body, and cannot be weighed together with it, not all matter is heavy.[30] The picture painted by Wolff presented an image of pieces of matter moving in different directions and meeting at the centre of the earth, where they would be obliged to block each other's path. Wolff responded to this problem by resorting to Descartes and Kepler and experiments which indicated "that if fluid matter moves in a circle around a central point, other matter being moved around it at a smaller speed is pushed towards the centre of the circle."[31]

By and large Wolff's concept of bodies and their motion is a mixture of metaphysical definitions of essence and (then prevalent) mechanistic ideas of the relation between body and motion, a mixture which also implies a conflation of impact and fall, elastic and non-elastic impact, throw, momentum and force, and which reveals that the "fluid matter" (so important in his debate with Keill) signifies both a profane, physical fluid and, at the same time, a metaphysical essence.

Wolff also deduces from this that a body has its force *within itself*. He writes: "Just as visible bodies, and all their parts, which we are able to discover with the best of our magnifying glasses, are not composed directly from the elements [more or less metaphysical entities], but always from even smaller bodies, so these bodies do not derive their force directly from the elements, but rather from fluid matters which flow through the space which is on occasion empty of peculiar matter. An example of this is heaviness, by means of which a body not only resists the motion of another, as we may witness for ourselves in water and other heavy fluid matters, but is also moved, as demonstrated by the fall of a stone from a height. ... [But as nothing occurs without reason, it is evident] that motion can only be increased by constant impacts; therefore the matter of the stone must, as it falls, be constantly receiving impacts towards the

27 Cf. ibid., pp. 121-124.
28 Cf. ibid., p. 125 f.
29 Cf. ibid., p. 131.
30 Cf. ibid., p. 129.
31 "Es zeiget demnach gegenwärtiger Versuch, daß, wenn eine flüßige Materie sich um einen Mittel-Punct im Circul herum beweget, eine andere Materie, die sich nicht so geschwinde wie sie herum bewegen lässet, gegen den Mittel-Punct des Circuls getrieben werde". (Ibid., p. 137, see also pp. 133-137.)

centre of the earth. It follows that there must be something which everywhere permeates it and can reach its parts; that is, a subtle fluid matter is required which is in permanent motion. ... To the extent that a body has alien matter (for so we will call the matter which flows through the space left free by the peculiar matter) of this kind within it, to this extent we can say that it has its force within it."[32]

Matter, Body and Motion – Newton's Interpretation

Newton's formulation of his law of inertia sounds similar to Wolff's, but in substance its meaning is very different, as is his concept of force, heaviness and the relationship between body and motion. Newton writes: "The *vis insita*, or innate force of matter, is a power of resisting, by which every body, as much as in it lies, continues in its present state, whether it be of rest, or of moving uniformly forwards in a right line."[33] Elsewhere he formulates his first law of motion: "Every body continues in its state of rest, or of uniform motion in a right line, unless it is compelled to change that state by forces impressed upon it."[34]

In Newtonian mechanics there is no caesura between rest and motion, nor, indeed, between a motionless object and an objectless motion. Rather, there is a distinction between accelerated and rectilinear uniform motion, the latter being defined as calibrated motion (in the sense of an etalon of motion). On the basis of this distinction, the category of all inertial systems constitutes the physical space which provides the background for describing the accelerated motion of physical matter (for instance, gravitational interaction) in dynamics. This indicates, in passing, that there was no

---

32 "Gleichwie aber die sichtbahren Cörper, ja alle ihre Theile, die wir durch die allerbesten Vergrösserungs-Gläser entdecken können, nicht unmittelbahr aus den Elementen (quasi metaphysischen Entitäten), sondern nur immer aus kleineren Cörpern zusammengesetzet sind; also haben ebenfalls dieselben Cörper nicht ihre Krafft unmittelbahr von den Elementen, sondern vielmehr von flüßigen Materien, die sich durch den von eigenthümlicher Materie hin und wieder leeren Raum bewegen. Ein Exempel hiervon giebet uns die Schweere, dadurch ein Cörper nicht allein der Bewegung des andern widerstehet, wie wir solches augenscheinlich im Wasser und anderen schweren flüßigen Materien sehen, sondern auch beweget wird, wie solches der Fall eines Steines von der Höhe ausweiset. ... (Doch da nichts ohne Grund geschieht, zeigt sich,) daß keine Bewegung stets vergrössert werden kan, als durch stetes Stossen; so muß die Materie des Steins, indem er herunter fället, beständig gegen den Mittel-Punct der Erde gestossen werden. Demnach muß etwas seyn, welches ihn überall durchdringet, und zu seinen Theilen kommen kan, das ist, man hat dazu eine subtile flüßige Materie nöthig, die in steter Bewegung ist. ... Insoweit nun ein Cörper dergleichen fremde Materie (denn so wollen wir diejenige nennen, die sich durch den von seiner eigenthümlichen Materie freyen Raum beweget) in sich hat; in so weit kan man sagen, daß er seine Kraft in sich hat." (Ch. Wolff, *Vernünfftige Gedancken Von Gott, ... [Deutsche Metaphysik]*, op. cit., p. 434 f.)

33 *Sir Isaac Newton's Mathematical Principles of Natural Philosophy and His System of the World.* Translated into English by Andrew Motte in 1729. The translations revised, and supplied with an historical and explanatory appendix, by Florian Cajori. Berkeley 1947, p. 2 (Definition III).

34 Ibid. p. 13 (Law I).

place in classical mechanics for the inactivity of matter, a concept often attributed to it but which does not even arise in relation to the principle of inertia.[35]

Force is then determined (in the second law) as something which can alter the calibration of a body's state of motion, that is, the state of rest *or* rectilinear uniform motion; in other words: acceleration. Newton did not see force – as most of his contemporaries, including Keill and Wolff, did – as momentum or impact. By equating force with momentum or impact, they not only overlooked the physical conceptualization by which forces, far from being rectilinear and uniform, are always equivalent to accelerated motions, that is, alterations to rectilinear and uniform motions; they also passed over an opportunity to give conceptual form to the ability of matter to act, in spite of the fact that they were at such great pains, Wolff among them, to grasp force as the potential of matter to act.[36] Therefore, the role played by the second derivatives, and thus by the second law of motion, in the conceptualization of matter's potential to act is not recognized.[37] Not even Kant and Hegel understood this point.

Inherent in the different versions which Wolff and Newton gave to the principle of inertia, therefore, are their different concepts of force and matter. At the time, the controversy over these two concepts blew up around the concept of gravitation, i. e., the force "gravitation", and this controversy revealed the transformation which certain categories were now undergoing.[38]

If Newton insists at various points in his work that gravitation is not a property essential to matter,[39] this must be seen in the light of the contemporary concept of mat-

---

35 In his early fragment "De gravitatione...," published posthumously, Newton attempted a demarcation between rest and motion. This fragment is primarily a critique of Cartesian physics which enabled Newton to conclude that the Cartesian approach offered no basis for a mechanics in which both heaven and earth had a place. It is not legitimate, therefore, to cite this fragment in support of the claim (as occasionally happens) that Newton's mechanics includes a concept of absolute rest. Nevertheless, this theory does leave scope for distinguishing between rest and motion in the sense of $v = 0$ and $v \neq 0$, but only within a selective framework of references, not within an absolute system, which only exists for acceleration and is called "absolute space".

36 On the relationship between Newton's three laws see: the editors' annotations and epilogue to *Ernst Mach, Die Mechanik in ihrer Entwicklung – historisch-kritisch dargestellt*, hg. und mit einem Anhang versehen von R. Wahsner and H.-H. v. Borzeszkowski, Berlin 1988, pp. 534-536, 600-604; *Physikalischer Dualismus und dialektischer Widerspruch. Studien zum physikalischen Bewegungsbegriff*, Darmstadt 1989, in particular pp. 31-39.

37 See, e.g., H.-H. v. Borzeszkowski und R. Wahsner, "Die Mechanisierung der Mechanik. Rezeption und Popularisierung der klassischen Mechanik durch Voltaire," in *Newton-Studien*, Berlin 1978, and subsequent papers of the authors.

38 Cf. H.-H. v. Borzeszkowski and R. Wahsner, *Newton und Voltaire. Zur Begründung und Interpretation der klassischen Mechanik*, Berlin 1980, pp. 13-72; *Die Wirklichkeit der Physik. Studien zu Idealität und Realität in einer messenden Wissenschaft*, Frankfurt/Main–Berlin–Berne–NewYork–Paris–Vienna 1992, in particular pp. 29-40, 50-56, 109-114; R. Wahsner, *Das Aktive und das Passive. Zur erkenntnistheoretischen Begründung der Physik durch den Atomismus – dargestellt an Newton und Kant*, Berlin 1981, pp. 14-19, 54-60, 115-130; "Nicht die Einzelheit herrscht in der Natur der Dinge. Zum Wissenschaftsprinzip des kollektiven Individuums," in R. Wahsner, *Prämissen physikalischer Erfahrung*, Berlin 1992.

39 Cf. I. Newton, "Letters to R. Bentley of 17 January and 25 February 1692-3," in *Isaac Newton's Papers and Letters on Natural Philosophy*, ed. by I. B. Cohen, Cambridge (Mass.) 1958, pp. 298, 303; *Sir Isaac Newton's Mathematical Principles of Natural Philosophy*, op. cit. pp. 400, 547.

ter. In Newton's day, the only features that were regarded as essential to matter and thus as physically real were those which could be attributed to the atom as such. Gravitation, however, decreases as the distance between masses increases, and in this sense it cannot be attributed to the smallest parts of bodies as such in the same way as hardness, extension, indivisibility, mobility or inertia. Nevertheless, Newton wished to show that gravitation was something physical and not simply a mathematical figment, and he sought long and hard for a way to derive it from the primary properties of atoms, at least as a secondary effect. Finally, in the 31st Question appended to his *Opticks*, he introduced the idea that not only the hitherto recognized primary qualities as passive principles can be attributed to atoms, but that they are also moved by active principles, which are realized by dynamic interaction, and that these principles form the things themselves.[40] These principles express something which is constituted by the mutual relationship of bodies, by their behavior towards each other. As bodies are only heavy towards each other, gravitation (like any other physical force) is dependent on the existence of more than one body or point of mass, and its existence cannot be seen in isolation from its action. Therefore, gravitation, and force in general, cannot be conceived as the property of a natural object in itself, but nor can it be conceived as existing separately from bodies, that is, merely as a relationship. It simply cannot be reduced to an impact or contact.

If one does not share Newton's concept of matter as an incarnation of individual atoms with their passive properties, but integrates this mutual relationship between bodies into the very concept of matter (thereby redefining the concept *matter*), one would have to conceive gravitation as inherent to matter. In actual fact, Newton did this in his physics, but instead of calling the thing "matter", he called it "nature".

If, on the other hand, one adopts a different concept of matter from the one explicitly offered by Newton, and takes matter to mean what Newton called "nature", and if one accepts his formulation that gravitation is not inherent to matter or an essential property of it, then one would be obliged to isolate the active principles from matter, and thereby divide matter from motion. By contrast, if one claims that gravitation is inherent to matter without understanding that this means altering the concept of matter, one is subjecting mechanics to a mechanistic misreading by attributing properties to isolated, individual atoms or bodies which they can only have in relationship to one another. – This complex state of affairs prompted many a misinterpretation of Newton's position.[41] One of these was Wolff's simplistic description of English opinion quoted above. Commentators also missed the fact that Newton himself did not at

---

40 Cf. I. Newton, *Opticks*, with a foreword by A. Einstein, an introduction by Sir Edward Whittaker, a preface by I. B. Cohen, Dover 1951, p. 401.

41 Cf., e.g., the article: H.-H. v. Borzeszkowski and R. Wahsner, "Voltaire's Newtonianism. A Bridge from English Empiricism to Cartesian Rationalism and Its Implications for the Concept of Mechanics in German Philosophy," Preprint 147 of the Max Planck Institute for the History of Science, Berlin 2000 [also: Appendix of the present volume, pp. 83-107]; Introduction to: *Voltaire, Elemente der Philosophie Newtons/ Verteidigung des Newtonianismus/ Die Metaphysik des Neuton*, edited and with an introduction and appendix by Renate Wahsner and Horst-Heino v. Borzeszkowski, Berlin 1997.

all want to provide metaphysical proof of the existence of vacuum. Quite the reverse: he moots the vacuum in order to test the validity of his $1/r^2$ law.

Wolff's Concept of Mechanics

The different positions taken by Newton and Wolff, and by correlation Keill and Wolff in the written controversy described above, are rooted in two different concepts of mechanics. It would not be true to say that Wolff sought to replace accurate measurement and deduction by metaphysical ramblings, nor was it his intention to invent a new mechanics. He hoped, rather, to provide a metaphysical substantiation for existing mechanics which would enable him to incorporate it into a uniform edifice of science.

In Wolff's eyes, the art of measurement has a hodogetic function for the entire cognitive process, and above all it helps to improve metaphysics. "For the art of measurement makes the human mind most perfect, teaching man to study wordly wisdom (*Weltweisheit*) and other forms and sciences more comprehensively, more thoroughly and more usefully, providing the painstaking scholar with an unexpected tool, and supplying the greatest benefit in life."[42] "The mathematical form of teaching", he argues in his preface to the *Anfangsgründe Aller Mathematischen Wissenschaften*, "demonstrates the correct use of reason, that is, how we may achieve clear, comprehensible and complete concepts, and how we may derive the other things from these without offending".[43] He saw it as an appropriate means to "free oneself from the harmful sway of the senses and imagination, thereby happily avoiding all error and prejudice".[44] It was beyond dispute, he argued, that mathematics provided the key to nature's hidden treasures. In fact, Wolff went so far as to claim that "most earthly hap-

---

42 "Denn die Meßkunst machet das menschliche Gemüth sehr vollkommen, lehret ihn die Weltweisheit und andere Arten und Wissenschafften weitläufftiger, gründlicher und nüzlicher durchgehen, gibt ihm zu einer gründlichen Gelehrsamkeit unverhoffte Hülfsmittel an die Hand, bringet auch den allergrösten Nuzen im Leben." (Ch. Wolff, "Wie die Meßkunst den menschlichen Verstand ausnehmend bessere (1713)," in *Werke*, op. cit., vol. 21.3. Gesammelte kleine philosophische Schriften III, Hildesheim–NewYork, 1981, p. 362.)

43 "Die Mathematische Lehr-Art giebt den rechten Gebrauch der Vernunfft zu erkennen, wie man nehmlich zu klaren, deutlichen und vollständigen Begriffen gelange, und daraus ohne Anstoß die übrigen Sachen herleite." (Ch. Wolff, *Der Anfangsgründe Aller Mathematischen Wissenschaften Erster Theil, Halle 1717*, Preface.)

44 "Derowegen ist kein gewisserer Weg zur Erkänntnis der Kräffte des menschlichen Verstandes zu gelangen, als wenn man mit Ernst die Mathematischen Wissenschaften treibet, und ... ist dieses zugleich das sicherste Mittel zu hurtigem Gebrauche der Vernunfft, (um) sich von der schädlichen Herrschafft der Sinnen und Imagination zu befreyen, das ist, alle Irrthümer und Vorurtheile glücklich zu vermeiden." (Ibid.) – Wolff also regarded the sway of senses and imagination as one reason for the inability of Leibniz and Clarke to reach agreement in their correspondence, as "Mr. Clarke still clings greatly to the senses and imagination, from which flow common prejudices, and these obstruct truth in metaphysics. It cannot be denied, therefore, that those who are not accustomed to see deep enough into the matter are more likely to applaud him than Mr. Leibniz, whose case calls upon us to see further." (Ch. Wolff, "Preface to the German edition of the correspondence, 1720," in *Der Leibniz-Clarke-Briefwechsel*, translated and edited by V. Schüller, Berlin 1991, p. 315 (Annexe).)

piness is constructed upon mathematics, and without it no Republic would be well designed".[45]

By building on clear mathematical concepts, metaphysics – so long trapped in darkness – could be perfected step by step. Wolff believed that this was possible even though metaphysical truths are so very different by nature from mathematical concepts.[46] To him, it was a necessary course of action also, because as soon as the comprehensible concepts were dissected sufficiently, one would arrive at the universal concepts explained by metaphysics, and would formulate propositions which could only be demonstrated from the universal concepts provided in metaphysics. That is why one should not expect to find the same certainties in the sciences as one found in mathematics until metaphysics had become comprehensible.[47] It is clear from this that in Wolff's view metaphysics is capable of something, for its part, which mathematics is not, for, according to him, in explaining nature mathematics alone is not enough, because without the assistance of a well-founded metaphysics it succumbs to those monstrosities of the imagination which hinder our recognition of nature.[48] This explains also Wolff's principal objection to Newtonian mechanics.

Wolff valued the art of measurement as a bridle which restrained imagination and the senses, thereby preventing them from "disrupting reflection".[49] He did not assume that there was nothing left to learn once one had grasped the fundamentals of measurement, but that these made it easier to understand the other sciences more accurately and more thoroughly. He maintained that we do not achieve certainty or correctness in natural science, and acquire sovereignty over the beasts unless we make use of the art of measurement in the scholarship of nature.[50]

It is because he holds the art of measurement in such esteem that Wolff attaches importance to mechanics, the most mathematical science of his age. Mechanics to him is the art of motion, a science of the art of creating a stronger or faster motion than a given force could accomplish of itself. That is why mechanics talked of machines, namely of machines "by means of which the motive force can either be better empowered to move a greater load than otherwise, or to carry out the motion more speedily than otherwise", although mechanics actually addresses all the laws of motion.[51]

---

[45] Ch. Wolff, *Der Anfangsgründe Aller Mathematischen Wissenschaften Erster Theil*, op. cit., Preface.

[46] Cf. Ch. Wolff, "Foreword to the German edition of the correspondence, 1720," op. cit., p. 312, 314.

[47] Cf. ibid., p. 305.

[48] Cf. Ch. Wolff, "Review of the original English edition of the Leibniz-Clarke correspondence for *Acta Eruditorum*, 1717," in *Der Leibniz-Clarke-Briefwechsel*, op. cit., p. 278.

[49] "... die Rechenkunst, Trigonometrie und Auflösungskunst, werden uns allgemeine Regeln an die Hand geben, nach welchen sich der Verstand in dem Erfinden richtet, und die Einbildungskrafft, nebst den Sinnen im Zaum gehalten wird, daß sie das Nachdenken nicht stöhren; ...". (Ch. Wolff, "Wie die Meßkunst den menschlichen Verstand ausnehmend bessere," op. cit., p. 374.)

[50] Cf. ibid., p. 369, 374, 377.

[51] "Die Bewegungs-Kunst (Mechanica) handelt zwar eigentlich von allen Gesetzen der Bewegung. ... Insgemein aber redet man in der Mechanik nur von den Maschinen, dadurch die bewegende Kraft entweder vermögender gemacht wird, eine grössere Last als sonst zu bewegen, oder die Bewegung geschwinder als sonst zu verrichten." (Ch. Wolff, *Anfangsgründe Der Bewegungs-Kunst Oder*

Wolff argues that there are immutable laws of nature, according to which all forces generate their motion when moving something, and according to which machines also function. "That is why we tend to call all effects mechanical which must of necessity occur as they do and in no other way, in keeping with nature's immutable laws of motion, because of the structure or constitution of things."[52] In this sense, not only the earth, plants, animals and the human body are machines, but also the whole world building. In designating them as such, the intention is to indicate "that motions in this great world construction are no less than all the alterations and effects which we perceive in plants, animals and, indeed, in the human body itself, which occur of necessity from their structure following the eternal natural laws of motion, the possibility of which can only be grasped by the human mind by considering these two things".[53] To Wolff, therefore, to philosophize mechanically meant to explain "the effects of nature and art from the structure of the things in accordance with the natural laws of motion, and elucidating how they are possible by means of the former and in accordance with the latter".[54]

Wolff regards Archimedes as the founder of mechanics (insofar as it concerns dead force), even if Archimedes only addressed motion generated by machines conceived by human art, rather than motion in nature.[55] Obviously this concept of mechanics determines the basic mechanical concepts which Wolff adopts into his system: he calls anything which induces motion "force", and anything which resists it "load". A force which truly results in motion is a "live force" (*lebendige Kraft*), and one which merely preserves the load or resists motion is a "dead force" (*tote Kraft*). Anything which enables a force to generate a useful motion (useful in the above sense) is known as a "machine".[56] The magnitude of the force is expressed as the product *quantity of matter* × *velocity*.

If we examine Wolff's mechanics and everything he says about the underlying concepts in his *Metaphysik* and *Kosmologie*, it is evident that his mechanics only embraces what Newton calls practical mechanics, essentially terrestrial mechanics, or

---

Mechanick, in *Der Anfangsgründe Aller Mathematischen Wissenschafften Anderer Theil*, Halle 1716, p. 229.)

52 "Daher pfleget man alle Würckungen Mechanisch zu nennen, die nach den unveränderlichen Bewegungs-Gesetzen der Natur aus der Structur oder Beschaffenheit der Dinge nothwendig so und nicht anders erfolgen." (Ibid., p. 231.)

53 "Sie geben nemlich durch diese Benennungen zu verstehen, daß die Bewegungen in dem grossen Welt-Gebäude nicht weniger als alle Veränderungen und Würckungen, die wir bey den Pflantzen, Thieren, ja in dem menschlichen Cörper selbst wahrnehmen, nach den ewigen Bewegungs-Gesetzen der Natur aus ihrer Structur nothwendig erfolgen, und also ihre Möglichkeit allein durch Erwegung dieser beyden Sachen von dem menschlichen Verstande begriffen werden kan." (Ibid.)

54 "Wenn nun jemand sich rühmen wil, daß er mechanisch philosophire, so muß er die Würckungen der Natur und Kunst nach den Bewegungs-Gesetzen der Natur aus der Structur der würckenden Dinge erklären und klärlich erweisen, wie sie nach jenen vermöge dieser möglich sind." (Ibid.; see also: Ch. Wolff, "Von dem richtigen Begriffe der mechanischen Philosophie" (1710), in *Werke*, op. cit., vol. 21.3. Gesammelte kleine philosophische Schriften III, op. cit., p. 730.)

55 Cf. ibid., p. 725.

56 Cf. e.g. Ch. Wolff, *Anfangsgründe Der Bewegungs-Kunst Oder Mechanick*, op. cit., pp. 229-231.

perhaps what the Ancients meant by tevcnh. However, this was not a theory of nature. Wolff, then, is passing something off as a theory of nature which does not have the categorical constitu tion or conceptual orientation to be one.[57] It is reflected in a thinking at a watershed between, on the one hand, the Ancient and Medieval thinking based on the concept of substance (*Substanzdenken*) and, on the other, the Modern thinking based on the concept of function (*Funktionsdenken*),[58] in which motion is still explained predominantly on the basis of bodies (or at least primarily in terms of their structure), and behavior on the basis of properties. Philosophy still shows no awareness of the typical modern-age (*neuzeitlich*) conception which understands motion itself as an object.[59]

The following passages are additional evidence:
"I have shown clearly in experiments that there is a particular matter in nature which creates warmth by its motion; ..."[60] "I can demonstrate from experiments that there is a par ticular magnetic matter in nature which circulates around and through the magnet and which derives its force from it – the magnet conveys its force to the iron ..."[61] – "And because this book of nature's laws has not yet achieved perfection, but is almost only at the beginning, there is no alternative but that the mechanical explanation for many alterations which occur in nature must today still remain concealed, although the structure of bodies has been examined often enough."[62] – "... so he must explain the effects of nature and art from the structure of the things which achieve these effects in accordance with the natural laws of motion ... ."[63] – "Accordingly I began by considering the essence and nature of bodies and the properties which they derive therefrom."[64] – "Hence heaviness is created from motion, and therefore there is

---

57 Cf.: H.-H. v. Borzeszkowski and R. Wahsner, "Die Natur technisch denken? Zur Synthese von τέχνη und φύσις in der Newtonschen Mechanik oder das Verhältnis von praktischer und theoretischer Mechanik in Newtons Physik," in *On the Culture of Morality. Practical Philosophy in the Posttraditional Society*, ed. by M. Weingarten, Berlin (in print) [also: Preprint 87 of the Max Planck Institute for the History of Science, Berlin 1998].

58 See footnote 94.

59 Just this aspect is not regarded in the otherwise profound book by John V. Burns, *Dynamism in the Cosmology of Christian Wolff. A Study in Pre-critical Rationalism*, NewYork 1965.

60 "Ich habe durch Versuche klärlich gezeiget, daß eine besondere Materie in der Natur ist, durch deren Bewegung die Wärme entstehet; ... " (Ch. Wolff, Ausführliche Nachricht von seinen eigenen Schrifften, op. cit., p. 478.)

61 "Ich erweise aus Versuchen, daß es eine besondere magnetische Materie in der Natur giebet, welche um und durch den Magneten circuliret, und davon er seine Krafft erhält, – der Magnet seine Krafft dem Eisen mittheilet ..." (Ibid., p. 485.)

62 "... so muß er die Würckungen der Natur und Kunst nach den Bewegungs-Gesetzen der Natur aus der Structur der würckenden Dinge erklären ... ." (Ch. Wolff, "Von dem richtigen Begriffe der mechanischen Philosophie," op. cit., p. 731.)

63 "Ich habe demnach zu erst von dem Wesen und der Natur der Cörper gehandelt und ihre daher rührende Eigenschafften erkläret." (Ch. Wolff, *Anfangsgründe Der Bewegungs-Kunst Oder Mechanick*, op. cit., p. 231; see also footnotes 52-54.)

64 "Daher entstehet die Schweere aus den Bewegungen, und also giebt es eine schweermachende (*gravifica*) Materie , welche aber, weil sie die Schweere wirket, selbst nicht schweer seyn kan: ... Und diese Materie kan, weil sie das fallen der schweeren Dinge verursachet, nicht zu der Menge der

a heavy-making (*gravifica*) matter which, however, since it causes the heaviness, cannot itself be heavy: ... And this matter, because it causes the heavy things to fall, cannot ... be reckoned among the quantity of matter which is moved, nor can it resist their fall."[65]

The Relationship Between Mathematics, Mechanics, Physics and Metaphysics

By elaborating his mechanics Newton founded physics in the modern-age (*neuzeitlich*) understanding, and in so doing it triggered the need to investigate what, in the terms of this new science, the *physical* means; and this has to be done in contrast to *natural philosophy* and *mathematics*. By implication, Newton had created an entirely new meaning for the concept of the physical, physical law and the relationship between bodies and motion. He had provided these concepts with a new categorical constitution: the body-motion question was no longer framed by the categorical relationship *thing – property*, but by the new categorical relationship *object – behavior*.[66, 67]

Newton's mechanics grew from a synthesis between experimental research into earthly mass (above all by Galileo) and theoretical astronomy (primarily formulated by Kepler). The law of gravitation, which was central to the new science, responded to a question much discussed at the time about whether bodies fell to the ground for the same reason that planets remained within their orbit. Following the Copernican principle, Newton's gravitational dynamics assumed the unity of natural law, so that the same natural laws were at play on earth as anywhere else in the universe, and he proved it. Nature now embraced heaven and earth, things and phenomena were universally comparable – an idea pursued since the dawn of the Modern Age, but now realized for the first time in a scientific theory.[68] Gravitation made this possible because it takes effect universally, over distances of both terrestrial and cosmic magnitude. That is why it was easy to regard the law of gravitation as the law of the world. In

---

bewegten Materie ... gezehlet werden, noch auch ihrem Fall Widerstand tun." (Ch. Wolff, *Ausführliche Nachricht von seinen eigenen Schrifften*, op. cit., p. 499.)

65 "Daher entstehet die Schweere aus den Bewegungen, und also giebt es eine schweermachende (gravifica) Materie , welche aber, weil sie die Schweere wirket, selbst nicht schweer seyn kan: ... Und diese Materie kan, weil sie das fallen der schweeren Dinge verursachet, nicht zu der Menge der bewegten Materie ... gezehlet werden, noch auch ihrem Fall Widerstand tun." (Ch. Wolff to J. Keill, quoted in the correspondence cited above, op. cit., p. 273.)

66 The term "object" translates here the German word "Gegenstand" determined within the framework of the philosophical conception "Gegenständlichkeit".

67 Cf. R. Wahsner, *Das Aktive und das Passive*, op. cit.; "Eigenschaft und Verhalten. Zur Beziehung von Mathematik und Physik," in *Gravitation und Kosmos. Beiträge zu Problemen der Allgemeinen Relativitätstheorie*, Berlin 1988; H.-H. v. Borzeszkowski and R. Wahsner, *Physikalischer Dualismus und dialektischer Widerspruch*, op. cit.; "Die Natur technisch denken?," op. cit., and also the literature quoted there.

68 Cf. R. Wahsner, *Mensch und Kosmos. Die copernicanische Wende*, Berlin 1978, "Introduction".

this context, unlike Newton, one often understood "world" not as the solar system or as the system of fixed stars but in a philosophical sense. Gravitation, then, had en exceptional role to play in grasping the new mechanics and in determining its relationship with mathematics and philosophy (both natural philosophy and metaphysics).

In his *Principia* Newton claims that he has described gravitation mathematically, but not yet explained it physically. In the definitions given in this work he writes that he uses "the words attraction, impulse, or propensity of any sort towards a centre, promiscuously, and indifferently, one for another; considering those forces not physically, but mathematically."[69] Although he was fully aware that the insights gain in physics could not be prised out of the mathematical structure in which they were expressed, he was nevertheless dissatisfied with this state of affairs and sought an additional – so he believed – physical explanation. His dissatisfaction was rooted in the fact that Newton, as the founder of theoretical physics, had created something new, but that naturally the full consequences could not be perceived at the moment of birth, not even the consequences for the relationship between physics and mathematics and for the concept of physical explanation. But because he introduced the concept of the active principle, a "physical explanation" was implicitly no longer the same thing as deriving an explanation from the primary properties of atoms. Instead, it meant reducing the phenomena to laws of nature grasping physical interactions. This had made it necessary to come up with a mathematical expression, but it had also established the ability of natural bodies to act. All in all, this was to create a new conception of physical reality,[70] but to begin with – one might say – it simply created an objectively new concept.

Wolff adopted a different approach from Newton to providing a philosophical determination of physical reality. In this, cosmology is to physics what ontology is to philosophy in general. The physicist's investigations do not target the realm of cosmology, but are confined to the corporeal realm. The task of physics is to derive phenomena from (simple) corporeal realities.[71] Wolff writes in his *Vernünftige Gedanken von den Wirkungen der Natur*: "In metaphysics we speak of the essence of things for and in themselves, without devoting our attention to the reality which a thing of this or the other essence has or might achieve. In physics, we consider the reality of those things which have a certain essence."[72]

While according a considerable role to experiment, he sees its value as supplying a provisional or supplementary explanation: "Although these things ... are demonstrated by mathematics, this does not prevent experiments from offering explanations.

[69] *Sir Isaac Newton's Mathematical Principles of Natural Philosophy*, op. cit., p. 5.

[70] Cf. footnote 38.

[71] On this relationship, cf. also: J. E. Erdmann, *Versuch einer wissenschaftlichen Darstellung der Geschichte der neuern Philosophie*, Faksimile-Neudruck der Ausgabe Leipzig 1834-1853 in sieben Bänden, Stuttgart 1982, vol. IV, pp. 287-313.

[72] "In der Metaphysick redet man von dem Wesen der Dinge vor und an sich selbst, ohne auf die Würcklichkeit mit acht zu haben, die ein Ding, welches dieses oder jenes Wesen hat, erreichen kan. In der Physick aber siehet man auf die Würcklichkeit der Dinge, die ein gewisses Wesen haben." (Ch. Wolff, *Vernünfftige Gedancken von der Natur und ihren Würckungen [Deutsche Physik]*, op. cit., p. 35.)

... For experiments are the trials which we conduct in order to ensure all the more firmly that our reflections have not been so very unfortunate."[73] "Everywhere we will find that the truths confirmed in experiments are required as the grounds for explaining what happens in nature."[74]

Wolff's picture of mechanics is founded on a conflation of mechanics as a physical science and a mechanistic world view, hence his resolve to found an overall science which will integrate all the disciplines,[75] but also a "superfluous" desire that metaphysics should do something (identify and substantiate links) which *physical* theory had already achieved, and which Wolff's age possessed (in the form of Newtonian mechanics), something which metaphysics cannot actually achieve, but which conceals its own task (and there is, indeed, one which remains for it to carry out). The fundamental and necessary differentiation between (natural) science and philosophy is effaced by all this. The famous correspondence between Leibniz and Clarke, and Wolff's reading of it, are excellent examples. Wolff primarily has recourse to Leibniz' argument "that divine rule over natural things consists in their very preservation, and should not be understood as ajnqrwpopaqw~"."[76] He fails here to recognize a real and fundamental problem posed by physics and mechanics and defends simultaneously a concept of god which has to be adopted for the sake of an independent philosophy which is not subjected to the dictates of theology.[77]

Wolffians and Newtonians

Clearly, Wolff's mechanical philosophy is not the same thing as Newton's mechanics. That is not because Wolff rejects this kind of mechanics, but because he fails to recognize it. He seeks to eliminate the resulting philosophical (and physical) flaws by means of his system, integrating mechanics into a edifice of science and providing a metaphysical, ontological and cosmological substantiation.

On the surface, the dispute between the Wolffians and the Newtonians focused on the hypothesis that motion originated in the world due to god's direct intervention and the doctrine that space was the *sensorium dei*. It is the latter which explains why Wolff adamantly discarded the idea of empty space. Both these issues reflected a deep-seated physical problem of epistemological significance, but they were expressed

---

73 "Es werden zwar diese Sachen ... in der Mathematick demonstriret: allein dieses hindert nicht, daß man sie nicht auch durch Versuche erklären kan. ... Denn die Versuche sind die Proben, die man anstellet um sich desto mehr zu versichern, daß man in seinem Nachdencken nicht unglücklich gewesen." (Ch. Wolff, *Ausführliche Nachricht von seinen eigenen Schrifften*, op. cit., p. 495.)

74 "Überall wird man finden, daß die in den Versuchen bestätigte Wahrheiten als Gründe gebraucht werden, dadurch dasjenige erkläret wird was in der Natur geschiehet." (Ibid., p. 498.)

75 Cf. also: H. Ostertag, *Der philosophische Gehalt des Wolff-Manteuffelschen Briefwechsels*, in Ch. Wolff, *Werke*, op. cit., III. Abteilung. Materialien und Dokumente, vol. 14, Hildesheim–NewYork 1980, p. 13.

76 Wolff quoting Leibniz in: "Review of the original English edition of the Leibniz-Clarke correspondence for *Acta Eruditorum,*" 1717, op. cit., p. 269.

77 Cf. footnote 38.

through such a distorting prism of theology and ideology that objective solutions proved impossible. Ultimately, it was simply a matter of which campproved strongest.[78]

The controversy between the Wolffians and the Newtonians was also the latters' battle to free natural science of philosophical patronage (of whatever complexion). To this end, even Leibniz supporters such as Lambert stood up for Newtonian mechanics, attempting to place it on a Leibnizian foundation or else blending the two bodies of thought, while Newtonians such as Euler offered alternative philosophical interpretations for what they saw as defects in Newton's mechanics (action at a distance, empty space), and physicists such as Maupertuis and, again, Euler, who accepted Newtonian mechanics, devised a new physical framework for it.

The new natural science was not a branch of metaphysics, and this was by no means to its detriment. However, it implied a completely new cognitive approach[79] which needed to be understood philosophically. Insofar as it ever has been, this did not happen until Classical German philosophy, notably Kant, recognized that there was an essential difference between natural science and philosophy in terms of epistemological status and point of departure.[80]

The Wolffians, meanwhile, wanted a metaphysical cognition of nature, ascribing secondary scientific significance to empirical, mathematical research, however much they extolled the virtues of experimentation and mathematics. In their opinion, physics which did not culminate in metaphysics did not even deserve the name "physics". The Newtonians, for their part, despised metaphysics, which they regarded as humbug.

Wolff's cardinal objection to Newtonian mechanics was that it was not philosophy, and that it was therefore, or perhaps in addition, open to exploitation by deists and sceptics. He expressed this fear graphically in his review of the English edition of the correspondence between Leibniz and Clarke, in his prologue to the German edition, and also in his letters to Manteuffel. In one of these he wrote: "A little piece has been sent to me from Paris from the Spatio against the Newtonians, specifically Mr. Gravesand and de Voltaire, in which the concept they use is described as extremely dangerous to theology, holding out the temptation of a new kind of Spinosimi spiritualis ..., just as abuse of the principiorum Cartesii gave birth to Spinosismum materialem. I cannot refute the author's claim. ... It is altogether an error of the English that they confuse the imaginaria, which are so useful in mathematics, with the realibus, which can certainly be distinguished from those of metaphysics and physics. ... But all these are the fruits of Hobbesio, whose materialism Locke insinuates using a pleasant vehiculo to those who claim to be perspicacious where others are concerned and yet

---

78 For more details of Euler's position and the full range of the debate: R. S. Calinger, "The Newtonian-Wolffian Controversy (1740-1759)," in *Journal of the History of Ideas 30*(1969), pp. 319-330; cf. also H. Ostertag, *Der philosophische Gehalt des Wolff-Manteuffelschen Briefwechsels*, op. cit., pp. 50-152.

79 Cf. E. Cassirer, *Substanzbegriff und Funktionsbegriff. Untersuchungen über die Grundfragen der Erkenntniskritik*, Darmstadt 1990; *Das Erkenntnisproblem in der Philosophie und Wissenschaft der neueren Zeit*, Darmstadt 1994 (Reprint), vols. I-IV, especially vol. II, pp. 396-582; see also H.-H. v. Borzeszkowski and R. Wahsner, "Die Natur technisch denken?," op. cit.

80 Cf. Footnote 18.

will not make the effort to use their intelligence soundly in containing their imagination and senses. In France philosophy is anyway in a very sorry state. Those who would go deep cling to Cartesio. Some adhere to Locke. And that is why scepticism and deism have the upper hand among great scholars in Paris."[81] On another occasion he remarks: "The foreigners ... are fascinated by Locke's principles and Newton's attractions, in a word no true philosophers."[82] Of Musschenbroek, Wolff comments that "his head is full of nought but attractionibus Newtonianis". "Those who hold truck with Newtonian philosophy, which in my opinion is a non ens, are so lofty because Newton's great name swells them up, and the freethinkers also believed they have surmounted the highest peaks of reason, since they cannot distinguish reason from an excrescence of their imaginations."[83] As to the "Newtonian" Maupertuis: "He would probably be of one mind with Algarotti, both in his view of religion, and in what is known by the simple-minded as Newtonian philosophy, which does not warrant the name of physics, let alone of all philosophy. Much as I esteem Newton for his advanced geometry, I cannot regard him as so much as a novice in philosophy, let alone a philosopher. ... Mr. Maupertuis is one of those who admire the so-called Newtonian philosophy, confusing philosophy with the cognitione Naturae mathematica, and not distinguishing this adequately from mere geometry, on which I have stated my views more fully in the fifth part of the Elementorum Matheseos. ... But the so-called Newtonians despise the Cartesians, and in Paris they currently believe that there is no other philosophy but the Newtonian and Cartesian varieties. Geometry and astronomy constitute the core of this activity. In Paris philosophy is in a gloomy condition. Maupertuis assumes that Newton understood all there is to philosophy better than any other can."[84] Wolff firmly believed that: "The Newtonians are haughty creatures and

---

81 "Es ist mir von Paris eine kleine Piece von dem Spatio gegen die Newtonianer, nahmentlich H. Gravesand und de Voltaire zugeschickt worden, darinnen man den Begriff derselben als höchst gefährlich für die Theologie erkläret, und der zu einer neuen Art eines Spinosismi spiritualis ... verleitet, gleichwie der Misbrauch der principiorum Cartesii den Spinosismum materialem zur Welt gebracht. Ich kan dem Autori hierinnen nicht unrecht geben. ... Es ist überhaupt ein Fehler, daß die Engelländer die imaginaria, welche in der Mathematick ihren großen Nutzen haben, von den realibus, welche von jenen in der Metaphysick und Physick wohl zu unterscheiden sind, confundiren. ... Es sind aber alles noch Früchte von dem Hobbesio, deßen Materialismus Locke unter einem angenehmen vehiculo einflößet denen, die für andere scharfsinnig seyn wollen, und doch nicht sich anstrengen mögen ihren Verstand recht zu gebrauchen und der Einbildungskrafft und den Sinnen wehe zu thun. In Franckreich ist die Philosophie auch noch in einem sehr schlechten Zustande. Die weit gehen wollen, hangen an dem Cartesio. Einige bleiben bey dem Locke. Und daher hat in Paris unter den größten Gelehrten der Scepticismus und Deismus die Oberhand." (Ch. Wolff, "Letter to E. Ch. von Manteuffel of 19 April 1739," in H. Ostertag, *Der philosophische Gehalt des Wolff-Manteuffelschen Briefwechsels*, op. cit., p. 60 f.)

82 "Die Ausländer ... sind mit Lockens principiis und Newtons attractionibus eingenommen, mit einem Worte keine rechte philosophi." (Ibid., p. 61.)

83 "Er (Musschenbroek) hat weiter nichts als den Kopff mit attractionibus Newtonianis voll. ... Die mit der Newtonianischen Philosophie schwanger gehen, die ich vor ein non ens halte, sind überhaupt hoch intoniret, weil sie der große Nahme Newtons aufgeblasen macht, und die Freydencker meinen auch den höchsten Gipfel der Vernunfft erreicht zu haben, da sie doch Vernunft und Misgeburt der Einbildungskrafft nicht unterscheiden können." (Ibid.)

despise all who do not blow their trumpet, and yet nobody with any inkling of it can recognize the so-called Newtonian philosophy as a philosophy at all. I surely struck the calf between the eyes when I demonstrated as much in the fifth part of my Elementorum Mathesos."[85]

Wolff's opinion of Euler was particularly low. While he accuses all Newtonians of wanting to turn Newton "by force after his death from a great geometrician into a great philosopher", in Euler's particular case he carps: "It occurs to me that a few years ago I was sent an excerpt from Euler's letters in which he testifies to his distaste for philosophy and admits openly that, had he attributed importance to philosophy, he would not have succeeded to the extent that he has and would not have made such progress with his calculating. And I am aware that those who hold his calculations in high esteem, seeking and obtaining their own glory in them, share the verdict that he does not have the slightest knowledge of philosophy and is quite unversed in methodo demonstrandi veterum, and his principal work, the Mechanica, displays both these things clearly, as one will find by reading merely the first of his definitiones with the attendant corollaries, and the first propositiones, where it is evident that he cannot tell his way about the principium rationis sufficientis."[86] Wolff is in no position to

---

84 "Mit dem Algarotti wird er wohl zusammenstimmen, sowohl in Ansehung der Religion, als der einfältig sogenannten Neutonianischen Philosophie, so nicht einmahl den Nahmen der Physick, geschweige denn der gantzen Philosophie verdienet. So hoch als ich den Newton in der höheren Geometrie halte; so kan ich ihn doch für keinen Anfänger in der Philosophie, geschweige für einen Philosophen halten. ... Mr. Maupertuis ist einer von denen, welche die sogenannte Newtonianische Philosophie bewundern, indem sie die Philosophie mit der cognitione Naturae mathematica confundiren, und diese nicht genug von der bloßen Geometrie unterscheiden, wovon ich mich deutlicher in dem fünften Theile der Elementorum Matheseos erkläret. ... Die sogenannten Newtonianer aber verachten die Cartesianer, und in Paris glaubet man zur Zeit, es gebe keine andere Philosophie, als die Newtonianische und Cartesianische. Geometrie und Astronomie ist das Hauptwerk, so man betreibet. In der Philosophie siehet es zu Paris noch gantz finster aus. Maupertuis nimmet also an, Newton habe in der Philosophie alles beßer eingesehen, als ein anderer einsehen kan." (Ibid., p. 61 f.)

85 "Die Newtonianer sind hochmüthige Creaturen und verachten alle, die nicht mit ihnen in ein Horn blasen, und gleichwohl kan niemand die sogenannte Newtonische Philosophie für Philosophie erkennen, der von dieser einen Begriff hat. Ich werde das Kalb in die Augen geschlagen haben, da ich solches in dem fünften Theile der Elementorum Matheseos gesetzt." (Ch. Wolff, "Letter to E. Ch. von Manteuffel of 19 April 1739," in H. Ostertag, *Der philosophische Gehalt des Wolff-Manteuffelschen Briefwechsels*, op. cit., p. 62.)

86 "Mir fället aber ein, daß mir vor ein paar Jahren ein Excerptum aus Eulers Brieffe communiciret worden, darinnen er seinen degoust vor der Philosophie bezeiget und offenhertzig gestehet, wenn er sich auf die Philosophie hätte legen wollen, würde er es nicht weit gebracht haben, und nicht so weit wie mit seinem calculiren kommen seyn. Und ist mir bekandt, daß auch diejenigen, die ihn wegen des calculirens hochachten, und zugleich darinnen ihren Ruhm suchen und erhalten haben, von ihm nicht anders urtheilen, als daß er in der Philosophie nicht das allergeringste verstehet und im methodo demonstrandi veterum gantz ungeübet sey, wie beides auch sein Hauptwerck; nemlich die Mechanica; klärlich zeiget, wenn einer nur die ersten definitiones mit den angehängten corollriis, und die ersten propositiones lieset, wo man auch finden wird, wie er sich in das principium rationis sufficientis nicht zu finden gewußt." (Ch. Wolff, "Letter to E. Ch. von Manteuffel of 6 November 1746," in H. Ostertag, *Der philosophische Gehalt des Wolff-Manteuffelschen Briefwechsels*, op. cit., p. 74.)

judge whether Euler's method is appropriate from the angle of mechanics, although he believes he is. In fact, the principle of sufficient reason has no part to play here. Philosophical method is not the yardstick of all science. Wolff's assessment of English newtonianism, then, blends a mechanistic misunderstanding of mechanics, for which Newton's supporters were themselves to blame, with his own view of what physics and mechanics ought to be: "There is nothing as simple-minded about the English as that they insist on having a Newtonian philosophy, which amounts to a few points of physics and leaves much wanting. For one thing because the English have respect for nobody who will not blow their trumpet and praise Newtonian philosophy above all else: there are also some among the French who adore this atrophied philosophy and despise all else. And Mr. Maupertuis ranks among them. And as the English have to date not recognized Mr. Euler as he no doubt deserves in his particular way, but rather despise him, he now seeks to flatter them, to gain their favor, and others at the Academy in Berlin who have not much grasp of philosophy also believe that they will acquire acknowledgement in England for the sometimes very paltry and sorry works in their cause if they will only worship their idol and despise Mr. von Leibniz. I would not be content to watch all this with indifference if Mr. Maupertuis did not enjoy such great esteem and such high office, enabling him to harm me by insinuating his hostile judgments."[87] Wolff complains that some German supporters of Newton wished to promote their own image "by doing down the greatest of their compatriots in order to elevate the foreigners". He laments: "It is true that the so-called Philosophia Newtoniana has met with much applause beyond England in France, Italy and Holland, to the detriment of the sciences and to the advantage of freethinking, threatening to topple religion and honesty. All these yarns will vanish even in England as quickly as they were spun. Only truth prevails, and brings us unflawed fruits."[88]

---

87 "Es ist von den Engelländern nichts einfältiger, als daß sie eine Philosophiam Newtoniam haben wollen, da es auf einige wenige Puncte in der Physick ankommet, dagegen noch vieles zu erinnern. Unterdeßen weil die Engelländer von niemanden etwas halten, als der mit ihnen in ein Horn bläset, und die Newtonsche Philosophie über alles erhebet; so finden sich auch einige unter den Frantzosen, die diese kurtze Philosophie lieben, und alles andere verachten. Und unter diese gehört auch H. Maupertuis. Und da die Engelländer bisher H. Eulern nicht aestimiren wollen, wie er auch wohl in seiner Art verdienet hätte, sondern vielmehr verachtet; so wil er auch nun ihnen flattiren, um ihre Gunst zu erwerben, und die andere bey der Academie in Berlin, die sich in der Philosophie nicht weit verstiegen, vermeinen auch, es sollte ihre zum Theil sehr magere und elende Sachen in ihrer Histoire in Engelland in Ansehung kommen, wenn sie ihren Abgott verehrten und den H. v. Leibnitz verachteten. Ich wollte dieses alles mit gleichgiltigen Augen ansehen, wenn H. Maupertuis nicht in großem Ansehen an hohem orte stünde, und mir zum Nachtheil wiedrige Urtheile insinuiren könnte." (Ch. Wolff, "Letter to E. Ch. von Manteuffel, date illegible," in H. Ostertag, *Der philosophische Gehalt des Wolff-Manteuffelschen Briefwechsels*, op. cit., p. 101.)

88 "Es ist wahr, die sogenannte Philosophia Newtoniana hat außer Engelland in Franckreich, Italien und Holland vielen Beyfall gefunden, zum Verderb der Wißenschafften und zu Beförderung der Freydenckerey, welcher der Religion und Redlichkeit den Umsturtz drohet. Allein dieses Spinnen-Gewebe wird selbst in Engelland wieder vergehen, wie es sich entsponnen. Die Wahrheit kan bloß bestehen, und bringet Früchte, die unverwerfflich sind." (Ch. Wolff, "Letter to E. Ch. von Manteuffel of 11 July 1748," in H. Ostertag, *Der philosophische Gehalt des Wolff-Manteuffelschen Briefwechsels*, op. cit., p. 147.)

If Wolff regarded Newtonian theory as the downfall of scholarship, it was because his own inability to perceive the specific quality of Newton's mechanics as modern-age (*neuzeitlich*) science had been reinforced by a faulty representation of this science by its very adherents. He disapproved of the trends he witnessed: "Scholars with sound knowledge are fast dwindling everywhere in number, and superficiality is gaining the upper hand, now that the so-called Newtonian philosophy is combined with French fluff, whereby pleasant words draw the listener's attention away from reason towards imagination, reaping applause for so influencing the appetite. It will be difficult to obtain talented members for the Academies of Science, and I would be happier if diligent individuals did not leave you [Petersburg] so soon. Mr. Euler, who might enjoy his no doubt wellearned fame in higher mathematics, now seeks with all his might to dominate in all the sciences, for which he was never intended, given that he so much lacks the first rudiments, and also the reading required for historical insight: in this he causes much harm to his own reputation, for there are few who have any concept of the reputation he deserves, and he has brought much shame on the Academy of Science in Berlin, as the dispute over the monads, which he initiated, clearly demonstrates."[89] He then complains that the President is a Frenchman who cannot read German and is unfamiliar with the state of German scholarship. Otherwise he would surely keep Mr. Euler under more effective control. Admittedly, he observes, the man knows no more than Euler about anything apart from mathematics, but he possesses "greater cleverness and courtesy".[90]

Wolff's judgment of Newton can be summed up in the view that Newton was a thoughtful mathematician whose great contribution to mechanics was beyond doubt – even if it was not as great as that of Leibniz – but that mathematics and mechanics are not enough to understand nature if they lack metaphysical substantiation.[91] He is

89 "Die Gelehrten, welche gründliche Erkäntnis lieben, nehmen an allen Orten sehr ab, und es nimmt ein seichtes Wesen überhand, nachdem man mit einer sogenannten Newtonischen Philosophie das Frantzösische Fladder-Wesen verknüpft, da man blosz durch angenehme Worte die Aufmerksamkeit von dem Verstande auf die Seite der Einbildungskraft zurücke ziehet, und durch ihren Einflusz in den Appetit den Beyfall erhält. Es wird also schwer fallen geschickte Mitglieder in Academien der Wiszenschafften zu erhalten; und wollte ich wünschen, dasz tüchtige Personen nicht so bald von Ihnen wieder zurücke gingen [d.i. von Petersburg]. Herr Euler, der seinen wohl verdienten Ruhm in der höheren Mathematick geniszen könnte; wil nun mit Macht in allen Wiszenschafften dominiren, darauf er sich doch niemahlen geleget, und da es ihm so wohl an den ersten Gründen, als an Belesenheit fehlet, die zu einer historischen Erkäntnis erfordert wird: wodurch er so wohl seinem eigenen Ruhme sehr schadet, indem wenige sind, die von dem ihm gebührenden Ruhme einen Begriff haben, als auch die Academie der Wiszenschafften zu Berlin in viele Schande bringet, wovon der von ihm erregte Monaden-Streit eine klare Probe ableget." (Ch. Wolff, "Letter to J. D. Schumacher of 6 May 1748," in Ch. Wolff, Gesammelte *Werke*, op. cit., I. Abt. Deutsche Schriften, vol. 16. Briefe von Christian Wolff aus den Jahren 1719-1753, Hildesheim–New York 1971, p. 142 f.)

90 Cf. ibid.

91 In an account of the treasures of mechanics, he writes of Newton: "Versed in the art of measurement and profound in his thinking, Newton demonstrated some outstanding truths in his mathematical foundations for a natural science (princip. Philosoph. natur. mathem.), concerning both forces towards the central point and forces away from it, and he used these teachings for a mechanical explanation of the universe based on the concept which Kepler ... devised for it, that is, the true concept." ["So hat auch der tiefsinnige Meßkundige Newton in seinen mathematischen Anfangsgründen

quite correct to maintain that Newtonian mechanics is not philosophy, and basically this resolves the bone of contention between the Wolffians and the Newtonians. (i) At the time, however, this was not evident for a number of reasons:

- The title of the work *Philosophiae Naturalis Principia Mathematica*, in which Newton outlined his new mechanics, seemed to suggest that he claimed the right for his mechanics to be regarded as philosophy.
- Both the English and the French Newtonians (with Voltaire at the forefront) presented the matter as though the physical theory "Newtonian mechanics" were a philosophy or world view.
- At the time the epistemological distinction between science and philosophy had not been established. The Leibniz-Newton debate drew attention to the need to do so. During the Enlightenment, philosophy was taken to mean all-embracing science.

In addition (ii), Wolff did not realize that his own mechanical philosophy did not meet the standards of a natural science. His claim that it did leaned somewhat on Leibniz' contribution to the new physics – not regarding that all in all Leibniz' conception was not appropriate to found this new physics.[92] Wolff's mechanics, as we have already seen, was merely practical mechanics, or what the Greeks called tevcnh. The need for a mechanics which would function as a natural theory was no better understood than the need for a theoretical physics, or rather, the latter was confused with metaphysical substantiation.

It was evident from these two considerations [(i) and (ii)] that there was a need for a philosophy which comprehended the new natural science, both recognizing the new status of this science compared with the natural science of the Ancients, and also acknowledging that it was a scientific branch of its own right (although not autarch). One step in this direction was the voltairization of mechanics – Voltaire's fusion of rationalism and empiricism and, specifically, newtonianism, which professed to be a world view. We must bear in mind, too, that resistance to newtonianism took the form of a struggle against materialism, scepticism and deism (or what was mistaken for

---

der Naturlehre (princip. Philosoph. natur. mathem.) sowohl von dem zum Mittelpunct hinzu, als auch von demselbigen hinweg treibenden Kräfften vortreffliche Wahrheiten erwiesen, und diese lehren zu einer mechanischen Erklärung des Weltgebäudes nach demjenigen Begriffe, welchen Kepler ... sich davon vorgestellet hatte, das ist, nach dem wahren Begriffe, gebrauchet." (Ch. Wolff, "Von dem richtigen Begriffe der mechanischen Philosophie," op. cit., p. 727 f.) Elsewhere he comments: However much he [Clarke] defends Newton, none of these arguments promote the latter's standing. In mathematics the celebrated Newton quite rightly enjoys a supreme reputation, but in explaining nature mathematics alone is not enough, because without the assistance of a well-founded metaphysics it succumbs to those monstrosities of the imagination which hinder our perception of nature. (Cf. Ch. Wolff, "Review of the original English edition of the correspondence between Leibniz and Clarke for *Acta Eruditorum* 1717," op. cit., p. 278.)

[92] For further details see: H.-H. v. Borzeszkowski and R. Wahsner, "Die Metaphysizierung der Physik," in *Denkweise und Weltanschauung*, ed. by H. Horstmann, Berlin 1981; *Physikalischer Dualismus und dialektischer Widerspruch*, op. cit., in particular pp. 39-49.

them), even if its rational core was, in fact, the defense of metaphysics, or philosophy, whose justification is given now as before.[93]

For all the mutual aversion which divided the Wolffians and the Newtonians, we should not overlook a number of shared tenets. Both camps attached great importance to mathematics. Although they differed in their understanding of its role within the cognitive process – and this very difference is one of the clearest issues dividing them – they concurred in ascribing exceptional significance to its contribution. Ultimately, a mathesis universalis proved impossible, and yet it was indispensable to think over mathematical conceptualization in order to grasping the new principle on which thought would be based. This new principle has been aptly described as *functional thought* (*Funktionsdenken*), by contradistinction to the *substantial thought* (*Substanzdenken*) of the Classical world and the Middle Ages. (This term "functional thought" does not signify mathematical thought; rather, "function" is used here as a general term for *order, relationship, collective unity.*[94]) For both camps, the importance of mathematics is not merely rooted in its epistemological role, but also in the great (practical) *use* ascribed to it. On both sides, the usefulness of a conception was a decisive criterion. The Enlightenment did not require science simply as theory, but for guidance in (moral and) technical affairs. Moreover, both schools of thought sought to protect science from theology, no matter whether they called the learning they extolled science or philosophy.[95]

Conclusion

All this considered, it would be fair to claim that Wolff's system (not only his "mechanical philosophy") helped to promote an awareness in German philosophy at the time that a new type of rationality was needed, that the rational approach should not be diluted, that it should at all events be stringently applied, but that it must be designed in a manner to attribute greater scope and significance to empirical investigation.

---

93 Cf. footnote 81.

94 It has been claimed that "taking the concept of function from its mathematical shadow world and placing it into tangible material reality ... has become one of the more popular short cuts for idealism." (Cf. A. Thalheimer, "Über einige Grundbegriffe der physikalischen Theorie der Relativität vom Gesichtspunkte des dialektischen Materialismus," in *Unter dem Banner des Marxismus*, vol. 1 (1925/26), p. 306.) But this assessment overlooks both the complicated epistemological status of Modern-Age natural science and misreads the philosophical concept of *matter* when it is constituted within a philosophical system which does not fall behind Classical German philosophy. While it cannot be denied that the chosen designation potentially implies that objects can be dissolved into relationships, nor that the man who coined the term was guilty of assuming this on occasion, it is also correct that substantiality *(Gegenständlichkeit)* should not be eliminated, but reconceived.

95 One might expect to see this assertion refuted somewhere in the Leibniz-Clarke correspondence or, indeed, the Newton-Leibniz debate as a whole, but it is not. Although the participants argued about god's abilities and His relationship with the world, the controversial issues were actually determined by whatever "independence thesis" was currently at stake. (For partial elucidation see: R. Wahsner and H.-H. v. Borzeszkowski, "Introduction" to: *Voltaire, Elemente der Philosophie Newtons/ Verteidigung des Newtonianismus/ Die Metaphysik des Neuton*, op. cit., pp. 37-53.)

Wolff himself only managed to do this by referring to experience or invoking its authority. He did not incorporate empirical experience methodically into a philosophical system, nor did he achieve its "metaphysical" substantiation.[96] This would have meant granting natural science as a branch essentially different from metaphysics, a task which fell to Kant. However, Wolff did reveal the need for new thinking. In this respect he (and others) opened German windows to the Newtonian spirit, and in this he bears affinity with Voltaire. But, unlike Voltaire, Wolff had no desire to replace the old metaphysics with physics, or to declare that metaphysics and speculation were superficial. His aim was to renew the old metaphysics by integrating "mechanical philosophy" on a sound basis.

---

96 His concept of unity was a major hurdle here. As Heine correctly observes, Wolff only saw *unity* as completeness. (Cf. H. Heine, *Zur Geschichte der Religion und Philosophie in Deutschland*, in *Heines Werke in fünf Bänden*, Weimar 1958, vol. 5, p. 77.)

Hegel's Interpretation of Classical Mechanics *

Horst-Heino v. Borzeszkowski

Although Hegel praised Newton's grounding of the infinitesimal calculus, he was extremely critical of his work as a physicist and denied that his mechanics constituted an advance on Keplerian astronomy. He regarded his mechanics as nothing more than a naive application of the infinitesimal calculus, an unwarranted contamination of it. What is more, one can find in his *Lectures on the History of Philosophy*[1] remarks on this aspect of Newton's work which in the second half of the nineteenth century gave rise to the opinion that Newton was "an absolutist and a foolish inductivist".

The most interesting question raised by Hegel's criticism of Newton's mechanics is that of the viewpoint from which his assessment of Newton's work might possibly be justified. In fact finding an answer to this is essential to any proper understanding of modern physics.

Since the fact that the proper understanding of classical mechanics is essential for the epistemological and physical understanding of modern physics is not generally accepted, I shall begin with a few remarks in order to establish its general plausibility. The main point is that although it is certainly the case that Newton's mechanics, including his theory of gravitation, have been partly superseded by quantum mechanics and Einstein's theory of relativity, Newton's mechanics still has to be regarded as the prototype of all physical theories. I am not referring here to the range of their validity but to the structure and character of their concepts – all physical theories are constructed according to the rules of classical mechanics. It might also be said that unlike natural philosophy, all physical theories have the same epistemological status as classical mechanics. This is even true of physical theories which do not relate to mechanical phenomena, and of theories in respect of which Newton's theory has been shown to be no more than an approximation, if one is considering the range of its validity. What is more, Newton's classical mechanics is not only the pattern for all physical theories, it also plays a central role in the modern structuring of physics, for it is an essential part of the theory of measure basic to all physics.[2] If we keep in mind this present-day function of classical Newtonian mechanics, it becomes clear that if Hegel's criticism of it were valid, it would also have to be regarded as a valid criticism of modern physics in general.

---

* Reprint from: *Hegel and Newtonianism*, International Archives of the History of Ideas, vol. 136, ed. by M.J. Petry, Dordrecht: Kluwer Academic Publishers, 1993, pp. 73-80.

[1] Cf. G.W.F. Hegel, *Vorlesungen über die Geschichte der Philosophie III,* in *Werke in 20 Bdn., auf der Grundlage der Werke von 1832-1845 neu edierte Ausgabe*, Frankfurt a.M. 1986, vol. 20, pp. 231-233 [see also: LHP. *Hegel's Lectures on the History of Philosophy*, transl. by E.S. Haldane, 3 vols., London–NewYork 1963, III.322-324; Jub. *Vorlesungen über die Geschichte der Philosophie. Jubiläumsausgabe*, ed. by H. Glockner, Stuttgart 1968, vols. 17, 18, 19, 19.446-449].

[2] Cf. H.-H. v. Borzeszkowski and R. Wahsner, *Physikalischer Dualismus und dialektischer Widerspruch. Studien zum physikalischen Bewegungsbegriff*, Darmstadt 1989.

Let it be said at the outset, that the conclusion of our argument will be as follows: as far as classical mechanics as a physical theory is concerned, Hegel's criticism provides no valid arguments for changing classical mechanics. Hegel was not talking about classical mechanics; whenever he thought he was talking about it, he was really talking about a philosophical interpretation of mechanics, that is to say, the philosophical system of *mechanism.* Since his criticism of mechanism is not really relevant to the physical discipline of classical mechanics, strictly speaking it is no concern of modern physics. Nevertheless, his criticism is neither trivial nor is it completely devoid of interest for modern physics. In view of the identifying of mechanics with mechanism taken for granted by Hegel, his polemic against Newton is in many respects a profound criticism of mechanism as a philosophical conception. And this criticism is useful and necessary, since mechanistic interpretations of physical theories which do away with these theories in order to produce a philosophy are always harmful, both to philosophy and to physics. It is such mechanistic interpretations which prevent us from clarifying the genuine relationship between philosophy and the sciences, and it is precisely this clarification which is necessary if philosophy is to be fruitful within natural science and vice versa. It must not be overlooked, however, that Hegel's profound criticism of mechanism was formulated by misusing Newton's name and by misinterpreting the physical discipline of mechanics. It has to be added, moreover, that his identification of mechanics with mechanism necessarily gives rise to a distortion of the physical discipline.[3] It is perhaps worth observing, that this is why attempts to find an anticipation of modern physics in certain parts of Hegel's criticism are unacceptable.

Hegel's understanding of mechanics has certain definite shortcomings. As long as he is considering certain curvilinear motions such as the orbits of the planets, that is to say, as long as he is considering pure kinematics, his line of argument often shows a deep understanding of what he is dealing with. When the physical laws of dynamics constituting the basis of the calculation of the orbits are touched upon, however, he soon shows that he is unable to follow Newton. Incidentally, this lack of understanding is also a fairly typical feature of pre-Hegelian mechanistic interpretations of mechanics.

This limitation becomes fully apparent when he is discussing the infinitesimal calculus or comparing Newton's work with Kepler's. We shall now make a few remarks on these two points, which are of course closely inter-related.[4]

In his *Logic,* he reflected philosophically on the significance of the infinitesimal calculus, and his insight into the nature of this new branch of mathematics was so profound that he was better able to counter attacks on it than the mathematicians who were actually making use of it. He notes that in the calculus, mathematics is operating with the qualitative forms of quantities, and that it is therefore going beyond ordinary comprehension, since, "quantum in so far as it is infinite is required to be thought as sublated, as something which is not a quantum but yet retains its quantita-

[3] Cf. R. Wahsner, "Naturwissenschaft zwischen Verstand und Vernunft," in *Vom Mute des Erkennens. Beiträge zur Philosophie G.W.F. Hegels,* ed. by M. Buhr and T. I. Oiserman, Berlin 1981.
[4] Cf. H.-H. v. Borzeszkowski and R. Wahsner, *Physikalischer Dualismus und dialektischer Widerspruch.*

tive character."[5] He shows that this mathematical infinite is derived from the Notion of the true infinite. He emphasizes the fact that infinite quantum is not a determination of quantity which might actually exist as a quantum, but that it has significance only as something standing in relation to something else, that is to say in relation to a zero which is outside the relationship. In other words, it is only the relationship of the infinite, indefinite quanta dx and dy, the differential quotient dx/dy which is determined. Both the limit of the spatial differences and the limit of the temporal differences constitute a zero. Since the infinite quantum is therefore determined only in its essential unity with its counterpart, it is a determination of quantity in a qualitative form.

The objection was raised against the differential calculus, that an intermediate position between being and nothing is an impossibility. The calculus mentioned above is based on this assumption, however, for it derives from the notion that the determinations of quantum are vanishing quantities, that is, that they are neither a quantum nor a nothing, but a mutual determination in respect of other quantities. The objection raised was therefore rejected by Hegel, who maintained that the unity of being and nothing is not a state but a disappearing as well as a becoming, only the middle or the unity itself constituting the truth of the matter: "The said infinite magnitudes, therefore, are not merely comparable, but they exist only as moments of comparison, that is, of the ratio."[6]

Although Hegel appreciated the importance of the method of the infinitesimal calculus as a procedure corresponding to the true nature of the infinite, he regarded it as a drawback that mathematics had not deduced this method from the Notion: "We may begin by remarking that the method of differential calculus shows on the face of it that it was not invented and constructed for its own sake."[7] It has been extracted from empirical propositions and has justified itself only by calling attention to its undoubted usefulness. He found the reason for this drawback in the fact that mathematics contradicts by means of the infinitesimal calculus the very method on which it is itself based as a science. Mathematics is essentially the science of operating with finite quantities. Calculation in respect of the infinite requires procedures that are clearly at odds with this. At one and the same time, procedures relevant to computation in respect of finite quantities are being used in connection with infinite quantities. This notionless procedure apparent in the differential calculus, shows that this kind of mathematics is incapable of dealing with qualitative differences, and such a calculus is therefore quite unsuitable for physics.

When Hegel maintains that the infinitesimal calculus of his time was not well-founded, he is of course quite right. One has to agree with him completely when he objects to basing the calculus on, "an increment from the force of gravity",[8] or the ar-

---

5 G.W.F. Hegel, *Wissenschaft der Logik. Erster Teil*, in *Werke*, vol. 5, p. 283 [see also: WL. *Wissenschaft der Logik*, ed. by G. Lasson, 2 vols., Hamburg 1932, 1.242; *Science of Logic*, transl. by A.V. Miller, NewYork 1969, p. 243].

6 Ibid., p. 297 [WL I.255; tr. Miller p. 255].

7 Ibid., p. 323 [WL I.279; tr. Miller p. 279].

8 Ibid., p. 319 [WL I.276; tr. Miller p. 272].

gument of the "unimportance of the difference".[9] After Hegel, however, the formalism of the infinitesimal calculus was provided with a justification in a mathematically self-consistent manner – it was given an arithmetical foundation by Cauchy, Weierstrass, Cantor, Dedekind and others. This foundation consists in the fact that the continuity of real numbers is proved either by establishing proof of the fact that every Dedekind section in the field of real numbers has a sectional number, or that every real fundamental sequence has a limit. Reducing the infinitesimal calculus to purely arithmetical concepts showed that the operations carried out by mathematics within this calculus do not contradict the nature of finite determinations and their relations, and are therefore in accordance with mathematics in general,[10] provided that, to speak in Hegelian terms, finite determinations are taken to be thought-determinations that remain in opposition to each other as well as to what is objective and to the absolute.[11]

This notional justification of the calculus cannot be regarded as Hegelian, however, since it does not yield the "mathematics of nature" which Hegel was looking for. As he himself pointed out, such a mathematics can only take over what is qualitative from experience, it cannot develop it out of itself.

This would be the point at which one would have to postulate the independence of physics while also acknowledging its inseparable connection with mathematics and philosophy. Hegel was of the opinion, however, that the philosophical Notion is capable of resolving this issue. Ignoring the fact that the philosophical Notion is only effective in so far as it is conditioned by the principle of scientific experience,[12] he maintained that an unmediated synthesis of mathematics and philosophy, a "philosophical mathematics" was possible, and that it would give rise to a "science of measures". He envisaged a *philosophical* science of mathematics, proceeding from the Notion, providing proof of the qualitative moments, and so overcoming the basic incapacity of ordinary mathematics. As a doctrine of quantities it would be the science of measures,[13] and would therefore provide a "mathematics of nature"[14] which would be a realization of the ideal of Hegelian physics. This view of the matter prevented him from recognizing the real importance of the infinitesimal calculus for grasping the nature of physical motion.

Hegel regarded the work that had then been done on this "science of measures" as preparatory. What was needed was a comprehension of the "empirical numbers of nature", an extension of the work of Keoler and Galilei who had endowed these num-

---

9 Ibid., p. 303, 306 f. [WL I.260, 264; tr. Miller pp. 259, 262].

10 Cf. ibid., p. 296 [WL I.254; tr. Miller p. 254].

11 Cf. G.W.F. Hegel, *Enzyklopädie der philosophischen Wissenschaften. Erster Teil. Die Wissenschaft der Logik*, in *Werke*, vol. 8, pp. 91 f. (§ 25) [see also: *Hegel's Logic*, transl. by W. Wallace, Oxford 1975, p. 45].

12 Cf. H.-H. v. Borzeszkowski and R. Wahsner, "Die Notwendigkeit der Philosophie für die Naturwissenschaft," in *Dialektik 1. Orientierungen der Philosophie*, ed. by B. Heidtmann, Cologne 1980.

13 Cf. G.W.F. Hegel, *Enzyklopädie der philosophischen Wissenschaften im Grundrisse (1830). Zweiter Teil. Die Naturphilosophie.* Mit den mündlichen Zusätzen, in *Werke*, vol. 9, pp. 52-54 (§ 259 Remark) [see also: *Hegel's Philosophy of Nature*, ed. and transl. by M.J. Petry, 3 vols., London–NewYork 1970, vol. I, 253; *Hegel's Philosophy of Nature*, transl. by A.V. Miller, Oxford 1970, p. 37].

14 G.W.F. Hegel, *Wissenschaft der Logik. Erster Teil*, op. cit., p. 406 [WL I.353; tr. Miller p. 343].

bers with the general form of determinations of quantum. It would therefore be necessary to elicit these laws from inter-related qualities, which in the case of the laws of Kepler and Galilei would be the qualities of space and time. This was, however, work that still had to be done.

Hegel therefore realized that the infinitesimal calculus goes beyond the bounds of a mathematics of counting and of considering the relations of numbers. He had the critical acumen to see that it still lacked what was required to make it a science, and he realized that ordinary mathematics was unable to supply this factor. He saw this cognitive discovery only in a negative light, however, and confined himself to pointing out that although the infinitesimal calculus provides possibilities reaching out beyond ordinary mathematics, it is not yet philosophy. This view is wholly consistent with his system, within which there is no room for an independent physics working as a natural science based on measurement and employing correct and independent methods, that is to say, methods which are not contestable philosophically.[15]

The consequence of this view is that the only conceivable alternative is a directly philosophical grasping of the existing contradiction, of the motion being investigated. The infinitesimal calculus as a purely mathematical means of grasping motion physically is not taken into consideration.

Consequently, Hegel always takes the differential quotient dx/dy as it is in itself, and relates it only to the infinite. He ignores the fact that it creates a new quantity, and that on account of the duality in space-time and velocity, it also creates the possibility of representing physically the fact that: "to move means ... to be in this place and not to be in it, at one and the same time".[16] Although Hegel was of course aware that in mechanics the infinitesimal calculus is applied to the representation of motion, he considered this to be a mere application, a contamination of the idea of the calculus. That is why, although he had a high opinion of Newton as a mathematician, he had no real understanding of his work in physics, and insisted that he had merely demonstrated mathematically what Kepler had discovered by experience.[17] In the above-mentioned section on the infinity of quantum, he observes that: "The thought cannot be more correctly determined than in the way *Newton* has stated it. I eliminate here those determinations which belong to the idea of motion and velocity ... because in them the thought does not appear in its proper abstraction but as concrete and mixed with non-essential forms."[18] After an exhaustive investigation, Hegel comes to the conclusion that, "the application of the differential calculus to the elementary equations of motion does not of itself offer any *real* interest; the formal interest comes from the general mechanism of the calculus".[19] It was therefore no accident that in the *Science of Logic* he devoted so many pages to Lagrange's *Theory of Analytical Functions* but

15 Cf. R. Wahsner, "Naturwissenschaft zwischen Verstand und Vernunft," op. cit.

16 Hegel G.W.F. Hegel, *Vorlesungen über die Geschichte der Philosophie I,* in *Werke*, vol. 18, p. 314 [LHP I.273; Jub. 17.337].

17 Cf. p. G.W.F. Hegel, *Wissenschaft der Logik. Erster Teil,* op. cit., p. 321 [WL I.277; tr. Miller p. 273].

18 Ibid., p. 298 [WL I.256; tr. Miller p. 255].

19 Ibid., p. 348 [WL I.301; tr. Miller p. 294].

never mentioned his *Analytical Mechanics.* The latter work was of decisive importance for the elaboration of classical mechanics and for the further development of physics, since on the basis of the differential calculus, of the calculus of variables created by Lagrange and of d'Alembert's principle, it endowed Newtonian mechanics with a form which made it possible to apply it to solids and liquids, and subsequently also to use it for non-mechanical physical systems such as electro-dynamics. For Hegel, the equations of mechanical motion formulated by Lagrange were of no more interest than those formulated in purely Newtonian mechanics.

The nature of Hegel's misunderstanding of mechanics and therefore of physics in general also becomes obvious in his comparing of the work of Kepler and Newton.[20] He denied that Newton had proved Kepler's laws, and believed that he had only reformulated the third of them. He was therefore of the opinion that Newton's fame was unjustified, and that by right it belonged to Kepler. He maintained that a proof of Kepler's laws required the establishment of a connection between the laws of falling bodies and those of planetary motion. He attempted to elicit this connection by considering a philosophical sequence of stages leading from a relatively free to an absolutely free motion. The truth of the matter is, however, that Newton had in fact already established this connection by means of his formulation of the dynamical laws of classical mechanics, or more precisely, by his theory of gravitation. Newton's laws connect together Galilei's law of free fall, Kepler's laws, and a multitude of other gravitational motions, because all of them can be calculated as different solutions from Newton's general theory. Hegel's view of mechanics was therefore the result of his having failed to grasp the significance of dynamics.

What Hegel understood as mechanics is what is included in his philosophical system within the first of the three spheres of nature. This sphere is sub-divided into a further three stages. At the second stage of mechanics, the relation between matter and motion is presented as spatial separation and relatedness. This section is introduced as follows: "Matter maintains itself against its self-identity and in a state of extrinsicality, through its moment of negativity, its abstract *singularization,* and it is this that constitutes the *repulsion* of matter. As these different singularities are one and the same, however, the negative unity of the juxtaposed being of this being-for-self is just as essential, and constitutes their *attraction,* or the continuity of matter. Matter is inseparable from both these moments, and constitutes their negative unity, that is, singularity. This is however still *distinct* from the *immediate* extrinsicality of matter, and is therefore *not* yet *posited* as being a *center,* a *material* singularity *of an ideal nature,* that is, *gravity.*"[21] Consequently, Hegel argues as follows: matter is the unity of the two moments of repulsion and attraction; it presents itself as weightedness, which is understood as the tendency toward the center of a distributed materiality, within which the center is something purely geometrical and not physical. Matter itself is weighted: the property of weightedness cannot be separated from it, and displays itself

---

[20] Cf. G.W.F. Hegel, *Enzyklopädie der philosophischen Wissenschaften im Grundrisse (1830). Zweiter Teil. Die Naturphilosophie*, op. cit., pp. 85-106 (§ 270) [tr. Petry I.263-281; tr. Miller pp. 65-83].

[21] Ibid., p. 60 f. (§ 262) [tr. Petry I.241; tr. Miller p. 44].

as the tendency toward a center lying outside matter. The center must not be assumed to be material, "for the precise nature of material being is that it posits its center as *external to itself*".[22] In accordance with his general Notion of nature, Hegel saw in this determination of weightedness evidence of matter's lack of independence.

The essence of all physical theories, however, including classical mechanics, consists in their determining the concept of physical bodies by describing their mutual interaction, the ways in which they act upon one another. It is precisely on account of this that physical entities are different from geometrical ones. Since physics itself therefore provides arguments in favor of the independence of nature or matter, its very existence constitutes a refutation of *mechanism.* This is also obvious from the fact that physics does not separate motion and body, does *not* maintain that for exterior, finite motion a principle holds true according to which a body at rest will continue as such through all eternity and a body in motion continue in the state indefinitely.[23] On the contrary, mechanics presupposes motion as a basic state, namely rectilinear uniform motion as being equivalent to rest. In physics, only positive and negative accelerations require a cause, or more precisely, a force. Solely on account of his mistaken assumption, Hegel takes it for granted that mechanics treats rest and motion in accordance with the principle of identity – motion is motion and rest is rest, and both are exterior to each other. His whole criticism of Newton and of mechanics in general is based on this assumption.

Newton's mechanics has many strong points which Hegel denied, and which he believed he had to develop out of his philosophical Notion. What he is criticizing, therefore, is not Newton's mechanics but the *mechanistic world-picture.* It is this criticism of mechanism which is justified, for it shows that in order to understand the natural world in its totality, one has to go beyond the singularized aspects of the separate natural sciences.

---

[22] Ibid., p. 62 (§ 262 Remark) [tr. Petry I.263-281; tr. Miller p. 46].

[23] Cf. ibid., p. 69 (§ 266 Remark) [tr. Petry I.249,35; tr. Miller p. 52].

The Philosophical Background to Hegel's Criticism of Newton *

Renate Wahsner

If one is intent on developing a natural philosophy which is fully adequate to present-day requirements, one cannot disregard either Newton or Hegel or the relationship between them. It is evident enough, moreover, that such a natural philosophy can only be developed by determining the significance of Newtonian physics for modern science, and the manner in which the system of Hegelian dialectics might now be constructively superseded. This latter task will involve our analyzing Hegel's natural philosophy in the light of the scientific and epistemological issues it raises. This, in its turn, brings us face to face with the problem of the relationship between Newton and Hegel, the philosophical background to Hegel's criticism of Newton, the question of the precise significance of his analysis of Newtonian mechanics.

Hegel may be said to have established not only that philosophy has to accord with experience, but also that natural philosophy, both in its origins and in its further development, necessarily presupposes empirical physics.[1] It may seem rather strange, therefore, that his own philosophy of nature should involve so much criticism of Newton. He certainly had no very high opinion of Newton's ability to deal with thoughts *[dem Denken der Gedanken]*. In his extensive and well worked out Lectures on the History of Philosophy, he devotes only two pages to Newton. In them, he notes that Newton had reminded the sciences that they should beware of thinking. He evidently believed that Newton was responsible for the fact that the physical sciences were simply not interested in examining their own presuppositions. He was quite clearly of the opinion that Newton was not aware that when he was speaking about forces he was actually dealing with notions, that he was in the habit of handling ideas as if they were material objects of the same kind as "stone and wood".[2] He hauled him over the coals for having, "flooded mechanics with a monstrous metaphysics".[3] He therefore took it to be Kant's great merit to have demonstrated that physics makes use of

---

* Reprint from: *Hegel and Newtonianism*, International Archives of the History of Ideas, vol. 136, ed. by M.J. Petry, Dordrecht 1993, pp. 81-90.

1 Cf. G.W.F. Hegel, *Enzyklopädie der philosophischen Wissenschaften im Grundrisse (1830). Zweiter Teil. Die Naturphilosophie. Mit den mündlichen Zusätzen*, in *Werke in 20 Bdn., auf der Grundlage der Werke von 1832-1845 neu edierte Ausgabe*, Frankfurt a.M. 1986, vol. 9, p. 15 (§ 246 Remark) [see also: *Hegel's Philosophy of Nature*, ed. and transl. by M.J. Petry, 3 vols., London–NewYork 1970, vol. I, 197,9; *Hegel's Philosophy of Nature*, transl. by A.V. Miller, Oxford 1970, p. 6].

2 Cf. G.W.F. Hegel, *Vorlesungen über die Geschichte der Philosophie III*, in: Werke, vol. 20, pp. 231-233 [see also: LHP. *Hegel's Lectures on the History of Philosophy*, transl. by E.S. Haldane, 3 vols., London–NewYork 1963, III.322-324; Jub. *Vorlesungen über die Geschichte der Philosophie. Jubiläumsausgabe*, ed. by H. Glockner, Stuttgart 1968, vols. 17, 18, 19, 19.446-449].

3 G.W.F. Hegel, *Enzyklopädie der philosophischen Wissenschaften. Zweiter Teil. Die Naturphilosophie*, op. cit., p. 88 (§ 270 Remark) [tr. Petry I.265,21; tr. Miller p. 67].

thought-determinations without undertaking any critical examination of them, despite the indisputable fact that they constitute such an essential part of its subject-matter.[4] By demonstrating this, Kant had called attention to the difference between philosophy and empirical science. According to Hegel, although not according to Kant, this difference consists in the fact that unlike philosophy, "the sciences can presuppose their subject-matters as immediately presented and provided, and their method of cognition as already accepted, both in respect of its source and in respect of its progress".[5]

As Hegel pointed out, however, since Kant did not clarify the source of the thought-determinations physics is using, he initiated a dialectic of natural philosophy but did not manage to elaborate it.[6] It was this task that Hegel had set himself. The question is, however, whether he did in fact manage to throw light on the epistemological presuppositions of physics and so reveal the source of the concepts it employs, and whether this did in fact enable him to correct what he regarded as the Newtonian aberration.[7]

The Epistemological Status of Sciences in Hegel's System

Hegel defined physics as not only a consideration of nature based on perception and having as its object the cognition of general orders and classes of things, but also a thinking consideration of nature. Natural philosophy, according to him, has as its object a more comprehensive treatment of the subject-matter of physics, more comprehensive in that it considers this subject-matter in its own immanent necessity, that is to say, in the light of the self-determination of the Notion.[8] As Hegel expresses it, natural philosophy is therefore the speculative science of the Idea in its otherness. Nature in this sense, as the Idea in its otherness, may be regarded as the externality of the Idea, and from the fact that it is so considered, as externalized self-consciousness so to speak, it follows that it does not contain in itself its "absolutely ultimate purpose".[9] Externality therefore constitutes the basic determination within which the Idea exists as nature. This is also the case in that extrinsicality, the pervasiveness of difference,

---

[4] G.W.F. Hegel, *Vorlesungen über die Geschichte der Philosophie III,* op. cit., p. 364 [LHP III.456; Jub. 19.587]. As a comment see: R. Wahsner, *Das Aktive und das Passive. Zur erkenntnistheoretischen Begründung der Physik durch den Atomismus – dargestellt an Newton und Kant,* Berlin 1981.

[5] G.W.F. Hegel, *Enzyklopädie der philosophischen Wissenschaften. Erster Teil. Die Wissenschaft der Logik,* in *Werke,* op. cit., vol. 8, p. 41 (§ 1) [see also: *Hegel's Logic,* transl. by W. Wallace, Oxford 1975, p. 3].

[6] Cf. G.W.F. Hegel, *Vorlesungen über die Geschichte der Philosophie III,* op. cit., p. 364 [LHP III.456; 470-474; Jub. 19.587, 600-610].

[7] Cf. R. Wahsner, "Naturwissenschaft zwischen Verstand und Vernunft," in *Vom Mute des Erkennens. Beiträge zur Philosophie G.W.F. Hegels,* ed. by M. Buhr and T. I. Oiserman, Berlin 1981.

[8] Cf. G.W.F. Hegel, *Enzyklopädie der philosophischen Wissenschaften. Zweiter Teil. Die Naturphilosophie,* op. cit., p. 15 (§ 246) [tr. Wallace p. 47].

[9] Cf. ibid. pp. 13, 24, 62 (§§ 245, 247, 262 Remark) [tr. Petry I.195,15; 205,7; 242,35; tr. Miller pp. 4, 13, 46].

the appearance of differences as mutually unrelated existences, is peculiar to nature.[10] In the Hegelian conception, what is sensual is also very largely a matter of mutual externality.

In arriving at this basic determination of nature, Hegel was reflecting in a rather distorted and distorting manner on the measuring and calculating procedures of the natural sciences. Starting from the conviction that it is impossible for philosophy to accomplish anything really worthwhile on the basis of mere presentation, he took the investigation of presentation, the analysis of it into its constituents, to be the one method constituting the indispensable introduction to all true philosophizing.[11] This analysis is a method of finite cognition which takes the differentiated being with which it is presented as an opposition or presupposition, as a prior form of cognition, the activity of which consists in dissolving the concrete nature of what is given, isolating its differences, and endowing them with the form of an abstract generality.[12]

Such a view is not, of course, a completely adequate assessment of the methodology of the measuring and calculating sciences, but it does involve the essentially valid insight that comparison through measuring is only possible in that things or moments are separated. Measurement presupposes not only the differentiation of the moments of a concrete noun, but also the separating out of its moments.[13]

In thus characterizing natural objects, Hegel is in fact describing the epistemological implications of measuring quantities. It might be objected, however, that the particular determinations of these quantities are not to be regarded as separate entities. In the Hegelian system, the quantities involved in measurement, which from an epistemological point of view are a means to cognition, are ontologized and treated as natural objects, that is to say as the objects of an overriding analytical cognition. What is more, the equalities in behavior constituting the substance or content of the quantities measured, are interpreted as being things. As a result, the natural world as determined by Hegel corresponds to the view of nature developed by mechanicism, the world-view of the mechanistically-minded popularizers of natural science.

Since mind cannot confine itself to analysis, however, nature cannot be presented simply as an agglomeration of the differences inherent within mutually indifferent existent things. Nature is a system of stages, and the dialectical Notion develops the one out of the other.[14] Considered more closely, it will be found to deploy itself within the general sequence of mechanics, physics and organics. These stages specify the

---

[10] Ibid., p. 31 (§ 249 Remark), see also p. 24 (§ 247) [tr. Petry I.207,35; tr. Miller p. 16].

[11] G.W.F. Hegel, *Phänomenologie des Geistes*, in *Werke*, vol. 3, p. 35-37 [PhG. Bamberg–Würzburg 1807, pp. I-XCI; Hegel's Phenomenology of Spirit, transl. by A.V. Miller, Oxford 1977, pp. 1-45].

[12] Cf. G.W.F. *Hegel, Enzyklopädie der philosophischen Wissenschaften. Erster Teil. Die Wissenschaft der Logik*, op. cit., p. 397 f. (§§ 226, 227) [tr. Miller pp. 284-285].

[13] Cf. H.-H. v. Borzeszkowski and R. Wahsner, "Did the Nonlinear Irreversible Thermodynamics Revolutionize the Classical Time Conception of Physics?" *Foundations of Physics*, vol. 14 (1984), 653-670; *Physikalischer Dualismus und dialektischer Widerspruch. Studien zum physikalischen Bewegungsbegriff*, Darmstadt 1989.

[14] Cf. G.W.F. Hegel, *Enzyklopädie der philosophischen Wissenschaften. Zweiter Teil. Die Naturphilosophie*, op. cit., p. 31 (§ 249) [tr. Petry I.212,3; tr. Miller p. 20].

transition from extrinsicality or the infinite isolation of all that is individualized, through levels at which reality is immanently determined in respect of its form and so appears in reflective relationships, to the third stage, at which nature involves the determination of living being or subjectivity.[15] The Philosophy of Nature as a whole therefore demonstrates the "necessity of the Notion", the manner in which nature approximates to mind. Hegel has to view nature in this way, since he fails to see that the natural sciences themselves and not only natural philosophy, make it evident that nature is not simply an agglomeration of unrelated differences, that the natural laws formulated by the natural sciences are not merely abstract generalities, the set-theoretical formations of a certain kind of taxonomy.

As Hegel understands them, the natural sciences do not display the "necessity of the Notion". They are assumed to have a quite different source in immediacies, in things discovered, in presuppositions. Their subject-matter and their methodology are clearly derived at random from presentation,[16] and it is therefore quite unjustified to treat them as philosophy or Notional development.

Hegel himself emphasizes the indispensability of experience in overcoming the metaphysical hiatus between thought and objectivity. It should be noted, however, that when he does so, he is not concerned with experience which is specifically scientific, but with its everyday variety. He takes the principle of experience to involve "belief or immediate knowledge, ... the revelation of what is external in what is internal". This is the only connecting link he sees between thinking and reality.[17] The basic reason for his blinkered estimation of the natural sciences, for his denigration of Newton, the founder of theoretical physics, lies in this failure to distinguish properly between everyday and scientific experience, to appreciate the cognitive necessity of abstracting from the data provided by scientific measurement, which is something quite different from ordinary comparison, to grasp the true importance of the role played by measurement in establishing the empirical basis of the natural sciences.[18]

As has already been observed, however, he did at least recognize the general necessity of this empirical basis and of philosophy's having a good grasp of it. His doctrine was that philosophy, like empiricism, has to concern itself with that which is, that it must make no attempt to take into consideration that which has yet to be. The truly important principle involved here is that what is true is that which is present in reality and therefore accessible to perception. Since empiricism is limited to what is finite, however, it is bound to deny the supersensible, or at least the cognition and determination of it. And yet the laws it formulates are certainly more than what is sensual, and

---

[15] Cf. ibid., p. 37 (§ 252) [tr. Petry I.217,20; tr. Miller p. 25].

[16] Cf. G.W.F. Hegel, *Enzyklopädie der philosophischen Wissenschaften. Erster Teil. Die Wissenschaft der Logik*, op. cit., pp. 41, 52 f. (§§ 1, 9) [tr. Wallace pp. 3, 13].

[17] Cf. R. Wahsner, "Naturwissenschaft zwischen Verstand und Vernunft," op. cit.

[18] Cf. H.-H. v. Borzeszkowski and R. Wahsner, "Die Notwendigkeit der Philosophie für die Naturwissenschaft," in *Dialektik 1. Orientierungen der Philosophie*, ed. by B. Heidtmann, Cologne 1980; "Noch einmal über das Bedürfnis der Naturwissenschaften nach Philosophie," in *Dialektik 5. Darwin und die Evolutionstheorie*, ed. by K. Bayertz, B. Heidtmann and H.-J. Rheinberger, Cologne 1982.

necessarily go beyond mere experience.[19] Hegel argues that although it is true that the empirical sciences recognize the sterling value of measure amid the unmanageable ocean of empirical particularities, and that it is this which enables them to bring forth the thought of that which actually exists,[20] they persist in regarding thinking as nothing more than an abstraction or formalized generalization. He does in fact manage to demonstrate that this is the fundamental illusion of empiricism, which makes use of metaphysical categories such as matter, force, universality and the finite in formulating a line of argument, without being aware of the metaphysics involved. It therefore employs such categories and the connections implicit within them in a totally naive and uncritical manner.[21]

Hegel's criticism of empiricism involved making it aware of the untenability and triviality of its epistemological position, and the unwarranted self-satisfaction of the image it had of itself. The thought-determinations it employs in dealing with reality constitute the common ground it has with philosophy proper, however, and Hegel was fully aware that for this reason alone, it warranted a certain amount of respect. It was, after all, these thought-determinations which enabled him to treat the natural sciences as the precondition of a philosophical consideration of nature.

Physics, however, unlike empiricism, does not find its subject-matter simply in presentation, and is unable to rest assured that its particular method of cognition is a generally accepted one. The mechanical notion of "force", for example, as defined in the three Newtonian laws by the fundamental equations of mechanics, has an entirely different meaning from any which it might have in the context of mere perception. Hegel argues that thinking the empirical world involves changing its empirical form and transforming it into a generality. Yet it is the empirical sciences themselves which do this, and it can hardly be argued that their accomplishment is based solely upon the "necessity of the Notion". Such generalization is apparent in physics itself in the reproducibility of scientific facts. In order to make such a repro-ducibility possible, a form of physical experience has had to be developed which is essentially different from everyday experience – a comprehensive use has had to be made of the experimental method.[22] It is by comprehending the nature of this method that one can reach an understanding of the true epistemological status and preconditions of physics.

Hegel's fundamental mistake lies, therefore, in his identifying empiricism with the natural sciences. It is this that gives rise to his distorted view of Newton, for he persists in regarding Newtonian physics as indistinguishable from Lockean empiricism – "such metaphysicizing empiricism", he writes, "is now the norm in the sciences".[23]

---

19 Cf. G.W.F. Hegel, *Enzyklopädie der philosophischen Wissenschaften. Erster Teil. Die Wissenschaft der Logik*, op. cit., pp. 107-109 (§ 38) [tr. Wallace p. 61].

20 Cf. ibid., p. 49 f. (§ 7) [tr. Wallace p. 10].

21 Cf. ibid., p. 108 f. (§ 38 Remark) [tr. Wallace p. 61].

22 Cf. H.-H. v. Borzeszkowski und R. Wahsner, "Zur Beziehung von experimenteller Methode und Raumbegriff," *Dt. Zs. f. Philosophie*, vol. 28 (1980), 689-696.

23 Cf. G.W.F. Hegel, *Vorlesungen über die Geschichte der Philosophie III*, op. cit., pp. 209, 223, 233 [LHP III.312-313, 322-324; Jub. 19.438-439, 446-449].

As a result of identifying the natural sciences with the philosophical system of empiricism, Hegel is forced to deny them any proper independence or separate significance. Within his scheme of thought, nature is not allowed any status of its own. Consequently, the sciences concerned with nature are also devoid of any independent status, corresponding as they do to the location and significance accorded to nature itself within the Hegelian system. It is the dialectics of Hegelian philosophy which take over the scene, and the physical sciences which it reduces to mere empiricism are accorded very short shrift.

The fact is, however, that Hegel did not render the work of Newton superfluous, and that he was unable to put anything of any great significance in the place of Newton's truly constructive achievements. It would be impossible to base a measuring and calculating natural science on the principles of the Hegelian system. Natural science, or more particularly physics, is still based on the principles of Newtonian mechanics. What is more, neither Hegel's philosophical system nor any other could possibly make natural science superfluous, any more than any conceivable development in natural science could ever give rise to the demise of philosophy.

The Keplerian Laws and Gravitation

It was therefore unavoidable that this identification of empirical science with empiricism should have given rise to Hegel's criticism of Newton. Since Hegel took the identity of mechanics and mechanicism for granted, his wide-ranging polemic against Newton often turns out to be an extremely effective criticism of mechanicism. On account of the miscomprehension it involves, however, it entails much distortion of Newtonian physics.

Hegel's discussion of the relation between Newton and Kepler is a good illustration of the way in which he manages to misunderstand the issues at stake.[24] He denies that Newton was the first to prove the Keplerian laws of planetary motion, and will not admit that Newton brought about any real progress in the physical sciences: "Kepler proved the laws in that he discovered the universal expression of the empirical data. It has subsequently become customary to speak as if Newton were the first to have discovered the proof of these laws. The credit for a discovery has seldom been denied a man with more unjustness."[25] Hegel believes that he in his turn can prove that Newton's law of gravitation is simply a transformation of the third of Kepler's laws.

---

[24] G.W.F. Hegel, *Enzyklopädie der philosophischen Wissenschaften. Zweiter Teil. Die Naturphilosophie*, op. cit., pp. 85-106 (§ 270) [tr. Petry I.263,4; tr. Miller p. 65]. One can only get a full grasp of the representative nature of this issue by taking into consideration Hegel's whole conception of physics. See: R. Wahsner, "Naturwissenschaft zwischen Verstand und Vernunft," op. cit.; H.-H. v. Borzeszkowski and R. Wahsner, *Physikalischer Dualismus und dialektischer Widerspruch*, op. cit., pp. 11-17, 149-164.

[25] G.W.F. Hegel, *Enzyklopädie der philosophischen Wissenschaften. Zweiter Teil. Die Naturphilosophie*, op. cit., p. 86 (§ 270) [tr. Petty I.263,25; tr. Miller p. 66].

This law states that the squares of the periods, T, are to each other, as the cubes of great semi-axes, A, are to each other, that is, $\frac{T_1^2}{T_2^2} = \frac{A_1^3}{A_2^3}$. In other words, the ratio, of the cubes of the great semi-axes to the squares of the periods is a constant. Hegel therefore begins with the formula $\frac{A^3}{T^2} = \text{constant}$ by writing it $\frac{A \cdot A^2}{T^2} = \text{constant}$, and assuming Newton's definition of gravitation as $\frac{A}{T^2}$, substitutes in order to get $\frac{A}{T^2} = \frac{\text{constant}}{A^2}$, which expresses, "this so-called gravity as working in inverse proportion to the squares of the distances". Hegel takes this to be the Newtonian law of gravitation. As is very well known, however, Newton's law for the motion of mass m_1 under the influence of mass m_2 has the form in a simplified scalar notation of $m_1 b = G\frac{m_1 \cdot m_2}{r^2}$, where G denotes the gravitational constant, m_1 and m_2 the masses, r the distance between them and b the acceleration. One has to admit that Hegel is right with regard to the dimensions, for it is possible to transform the Newtonian law of gravitation into the form $b = \frac{G\frac{m_1 \cdot m_2}{m_1}}{r^2}$ and to posit the numerator on the right side as a constant. One then gets $b = \frac{\text{constant}}{r^2}$. This result corresponds to the Hegelian one in so far as r as well as A is a distance and the dimension of acceleration is the distance divided by the time squared. But this is as far as the matter goes, for the magnitudes in the two cases have different meanings. More particularly, the acceleration on the left-hand side of Newtonian law of gravitation is not equal to the left-hand side of Hegel's formula; consequently, it is not equal to the ratio of the mean distance to the square of the period T.

It is evident from this that for Hegel the essence of gravitation is to be found in the law of falling bodies. For him, free fall is characterized by s/t^2, that is to say, by the path divided by the square of the time. And this is precisely the dimension denoted above by the expression A/T^2 as the general Newtonian law of gravitation. Hegel wanted to bring out a connection between the law of falling bodies and the law of gravitation by a natural-philosophical development in which a progression is made from relatively to absolutely free motion. He did not see that this connection had been provided by classical mechanics.

The essential ingenuity of the Newtonian theory consists in its providing a law from which one can deduce in two different procedures, both the Keplerian laws of planetary motion and the Galilean law of falling bodies. It was this that had brought about the lawful and scientific unity of earth and cosmos. What is more, the theory also provided a general principle in accordance with which one could treat very different and perhaps previously unknown phenomena in a wholly consistent manner, providing further solutions to the same Newtonian equations, perhaps by adding a term designating a new force. It was certainly not the case that "the law therefore became increasingly superficial".[26] Of course if it were true that gravitation exists only as the law of falling bodies, Newtonian mechanics would indeed be superfluous.

Hegel overlooked the advantages of Newton's theory because what he had in mind was the pure deduction of the Keplerian laws from the Notion. If one translates the term "Notion" into the term "physical theory", the result is by no means unreasonable, for it provides us with a deduction of the Keplerian laws from Newtonian dynamics. It was precisely this, however, which Hegel considered to be superfluous and pointless.

Taken all in all, one might perhaps be able to take a positive attitude to this dialectic of the Notion by treating it in a figurative sense and bearing in mind what Hegel himself said of Kepler: that one should not laugh at his attempt to grasp the ordering of the solar system in accordance with the laws of musical harmony, for to do so would be to fail, "to appreciate the deep faith he had in the inherent rationality of this system; a faith which was the sole foundation of the brilliant discoveries made by this extraordinary man".[27]

Natural Science and Absolute Spirit

It should now be apparent that although Hegel's criticism is wholly justified in respect of mechanicism, metaphysics, empiricism and the philosophy of reflection, it is certainly wide of the mark in respect of Newton and the type of mechanics he initiated.[28] In assessing it, it is therefore essential not to overlook the fact that physics is not philosophy. Hegel's crucial mistake was that he allowed his systematic thinking to lead him into treating natural science as an essentially philosophical discipline. The rational element at the heart of his treatment of natural science is the insight that it is unable of its own accord fully to grasp it own presuppositions. It requires philosophy in order to do so. The truth of the matter is, therefore, not only that the empirical sciences are one of the major presuppositions of philosophy, but also that philosophy is arguably the most important presupposition of the sciences. Natural science and philosophy need one another. It is not the case, moreover, that philosophy might ever be able to dispense with the sciences, for they must always remain one of its premises. Although philosophy takes up and transforms the findings of the natural sciences, it does so not in a single procedure but by a process of perpetual assimilation. This is, moreover, a process which necessarily takes place at various levels, since the sciences change in time and complexity. They may lose their rapport with a particular philosophical framework, but they can never lose their rapport with philosophy as a whole.

Clarifying the origin of the thought-determinations employed by physics does not require that physics itself has to be reformulated. The clarification simply involves

---

[26] G.W.F. Hegel, *Phänomenologie des Geistes*, op. cit., p. 121 [PhG 59-100; tr. Miller pp. 79-103].

[27] G.W.F. Hegel, *Enzyklopädie der philosophischen Wissenschaften. Zweiter Teil. Die Naturphilosophie*, p. 131 (§ 280 Remark) [tr. Petry II.30,28; tr. Miller p. 103].

[28] Cf. H.-H. v. Borzeszkowski and R. Wahsner, *Newton und Voltaire. Zur Begründung und Interpretation der klassischen Mechanik*, Berlin 1980.

recognition of the categories being employed, the object being that their epistemological status may be ascertained, that an answer may be given to the old Kantian question of how science is possible. This will certainly involve the investigation of the basic characteristics of the experimental method and of scientific experience as such. The main outcome of it is already clear: the natural or physical sciences themselves are in a constant process of realizing their own "thinking attitude to objectivity".

It should not be forgotten, moreover, that the relationship in which the natural sciences stand to philosophy is by no means the only one which provides them with a meaning and a significance. They also have a function as the theoretical basis of productive technology and in making an essential contribution to the effectiveness of human labor.

On account of his having judged the natural sciences by purely philosophical standards, Hegel managed to identify their empirical aspect with metaphysicizing empiricism. He saw that this kind of metaphysics, in its turn, is conditioned by the fact that what goes by the name of an "object" is really a means. To a certain extent he was justified in regarding such objects as means, for within his system natural objects have their place between the primary and the concluding development of the stages of mind. As has already been pointed out, however, there are no such things as the real objects of nature, in the sense of a genuine counterpole to the mind.[29] Strictly speaking, therefore, there is nothing in nature which has to be mediated.

This identification of natural science with empiricism, together with the conclusions Hegel draws from it, reflect a real difficulty, however, for from a physical point of view quantities of measurement are cognized objects, while from an epistemological point of view they are a means to cognition. Quantities of measurement are, indeed, the objects of physics. Precisely considered, however, the natural science of physics does not make statements about real objects in the immediacy of their interactions, but about physical quantities and their relations. These quantities are therefore the means by which this reality is recognized. They are finite determinations, and if they were not, one would be unable to measure by means of them. Natural science cannot avoid operating by means of these quantities brought forth by the understanding. In doing so, however, it is dealing not with a metaphysics but with a physics founded theoretically in mensuration, and this is precisely what Newton's mechanics is.

The epistemological status of quantities of measurement is not of course to be identified with the Notion itself, the starting point of consciousness. The origin of such quantities has to be clarified, therefore, and there would appear to be no good reason for deriving them from presentation or everyday experience. Hegel assumed that this was their origin, however, and his nature, consisting as it does of the succession of stages, has this as its starting point. His whole conception of nature has therefore to be regarded as the ontologization of the mind's development from metaphysicized mechanicism to reflecting philosophy. This overcoming of the mind's externalization is therefore the overcoming of an ontologized mechanistic and reflective man-

---

[29] Cf. L. Feuerbach, "Zur Kritik der Hegelschen Philosophie," in *Werke*, ed by W. Schuffenhauer, vol. 9, Berlin 1970, pp. 40-41.

ner of thinking. In its negative aspect, therefore, that is to say, in so far as it is critical of metaphysics, empiricism and the philosophy of reflection, the process by which this overcoming takes place is a wholly valid procedure. In its positive aspect, however, it only has a partial validity, for Hegel fails to solve the problem of the origin of physical determinations of thought and physical objects. Without his general manner of philosophizing, however, the problem is indeed insoluble.**

** Thanks to M. J. Petry for pointing out a possible source of misunderstanding in the argument concerning Hegel's view of the relation between Kepler and Newton. The following additional remark may help clarification.

The question is whether Newton's law of gravitation can be derived from Kepler's laws, or more precisely, from his third law (in the mathematical or physical sense of 'derivation,' not, of course, in Hegel's philosophical sense).

Surely, without his knowledge of Kepler's laws, Newton never would have been in a state to develop the law of gravitation, nor would he have looked for it. After all, he was seeking to establish whether the same law holds responsible for falling bodies and planetary motion, so that the kinematic laws of planetary motion and the law of falling bodies (as well as the motion of the moon and the tide) would prove to be special cases of this universal law. Kepler's laws functioned, then, as an important prerequisite for Newton's derivation – but so did terrestrial mechanics (the law of falling bodies and Huygens' conception of central forces) and Newtonian axiomatics, the latter building on these astronomical-kinematic, and terrestrial-mechanical insights. By Newtonian axiomatics is meant the whole body of Newton's three axioms inclusive of the quantities contained therein and determined thereby.

For the crucial point why the law of gravitation cannot be *derived* from Kepler's laws lies in the fact that this law cannot be reduced to the term $G\frac{m_1 \cdot m_2}{r^2}$, but of necessity involves the complete equation given by the second law *mass* × *acceleration* = *force*. The force term $G\frac{m_1 \cdot m_2}{r^2}$ is substituted for *force* (that served as its place holder) on the right side of the equation and then this axiom is combined with the first and third axiom according to the rules. (Thus it might be added that the subject matter of Hegel's derivation is, in fact, *not* the law of gravitation.)

Once this is clear, we are free to admit that it is possible, through the above-mentioned presuppositions, to derive from Kepler's planetary laws the proof that the force term in question will be of the *form* of an $1/r^2$-law. The gravitational constant G has to be determined through other means. This mode of derivation is indeed part of elementary physics, and, moreover, historical investigations have also shown that belonged already to Newton's repertoire (see, for instance, A. R. Hall, "Newton on the Calculation of Central Forces," *Annals of Science,* vol. 13 (1957), 62-71; J. Herivel, *The Background to Newton's Principia. A Study of Newton's Dynamical Researches in the Years 1664-84,* Oxford 1965).

Index